PE Civil
Quick Reference
Sixteenth Edition

Michael R. Lindeburg, PE

PPI®

PPI2PASS.COM
A **KAPLAN** COMPANY

Register Your Book at ppi2pass.com

- Receive the latest exam news.
- Obtain exclusive exam tips and strategies.
- Receive special discounts.

Report Errors for This Book

PPI is grateful to every reader who notifies us of a possible error. Your feedback allows us to improve the quality and accuracy of our products. Report errata at **ppi2pass.com**.

PE CIVIL QUICK REFERENCE
Sixteenth Edition

Current release of this edition: 2

Release History

date	edition number	revision number	update
May 2018	16	1	New edition. Code updates. Copyright update.
May 2019	16	2	Minor cover updates.

PPI
1250 Fifth Avenue, Belmont, CA 94002
(650) 593-9119
ppi2pass.com

ISBN: 978–1–59126–573–3

Library of Congress Control Number: 2018936669

Table of Contents

How To Use This Book v

Codes and Standards vii

For Instant Recall

Fundamental and Physical Constants 1
Temperature Conversions 2
SI Prefixes 2
Equivalent Units of Derived and Common SI Units 2
Quick Engineering Conversions 3
Mensuration of Two-Dimensional Areas 5
Mensuration of Three-Dimensional Volumes 7

Background and Support

CERM Chapter 3: Algebra 9
CERM Chapter 4: Linear Algebra 9
CERM Chapter 5: Vectors 9
CERM Chapter 6: Trigonometry 10
CERM Chapter 7: Analytic Geometry 12
CERM Chapter 10: Differential Equations 13
CERM Chapter 11: Probability and Statistical Analysis of Data 13
CERM Chapter 13: Energy, Work, and Power 16

Water Resources

CERM Chapter 14: Fluid Properties 19
CERM Chapter 15: Fluid Statics 20
CERM Chapter 16: Fluid Flow Parameters 21
CERM Chapter 17: Fluid Dynamics 22
CERM Chapter 18: Hydraulic Machines 26
CERM Chapter 19: Open Channel Flow 28
CERM Chapter 20: Meteorology, Climatology, and Hydrology 31
CERM Chapter 21: Groundwater 32
CERM Chapter 22: Inorganic Chemistry 33
CERM Chapter 24: Combustion and Incineration 34
CERM Chapter 25: Water Supply Quality and Testing 37
CERM Chapter 26: Water Supply Treatment and Distribution 38

Environmental

CERM Chapter 28: Wastewater Quantity and Quality 41
CERM Chapter 29: Wastewater Treatment: Equipment and Processes 42
CERM Chapter 30: Activated Sludge and Sludge Processing 43
CERM Chapter 31: Municipal Solid Waste 46
CERM Chapter 32: Pollutants in the Environment 47
CERM Chapter 34: Environmental Remediation 47

Geotechnical

CERM Chapter 35: Soil Properties and Testing 49
CERM Chapter 36: Shallow Foundations 51
CERM Chapter 37: Rigid Retaining Walls 53

CERM Chapter 38: Piles and Deep Foundations 56
CERM Chapter 39: Excavations. 57
CERM Chapter 40: Special Soil Topics 59

Structural

CERM Chapter 41: Determinate Statics 63
CERM Chapter 42: Properties of Areas. 64
CERM Chapter 43: Material Testing 65
CERM Chapter 44: Strength of Materials 67
CERM Chapter 45: Basic Elements of Design 69
CERM Chapter 46: Structural Analysis I 73
CERM Chapter 47: Structural Analysis II. 74
CERM Chapter 48: Properties of Concrete and Reinforcing Steel 74
CERM Chapter 49: Concrete Proportioning, Mixing, and Placing. 75
CERM Chapter 50: Reinforced Concrete: Beams 76
CERM Chapter 51: Reinforced Concrete: Slabs 81
CERM Chapter 52: Reinforced Concrete: Short Columns. 82
CERM Chapter 53: Reinforced Concrete: Long Columns 82
CERM Chapter 54: Reinforced Concrete: Walls and Retaining Walls 83
CERM Chapter 55: Reinforced Concrete: Footings 84
CERM Chapter 56: Prestressed Concrete 85
CERM Chapter 57: Composite Concrete and Steel Bridge Girders 86
CERM Chapter 58: Structural Steel: Introduction 87
CERM Chapter 59: Structural Steel: Beams 87
CERM Chapter 60: Structural Steel: Tension Members 90
CERM Chapter 61: Structural Steel: Compression Members. 91
CERM Chapter 62: Structural Steel: Beam-Columns 94
CERM Chapter 63: Structural Steel: Built-Up Sections 94
CERM Chapter 64: Structural Steel: Composite Beams 96
CERM Chapter 65: Structural Steel: Connectors. 96
CERM Chapter 66: Structural Steel: Welding 97
CERM Chapter 67: Properties of Masonry 98
CERM Chapter 68: Masonry Walls. 98
CERM Chapter 69: Masonry Columns 100

Transportation

CERM Chapter 70: Properties of Solid Bodies. 101
CERM Chapter 71: Kinematics 101
CERM Chapter 72: Kinetics. 102
CERM Chapter 73: Roads and Highways: Capacity Analysis 104
CERM Chapter 75: Highway Safety 106
CERM Chapter 76: Flexible Pavement Design 107
CERM Chapter 77: Rigid Pavement Design 109
CERM Chapter 78: Plane Surveying. 109
CERM Chapter 79: Horizontal, Compound, Vertical, and Spiral Curves 111

Construction

CERM Chapter 80: Construction Earthwork. 117
CERM Chapter 81: Construction Staking and Layout 118
CERM Chapter 82: Building Codes and Materials Testing. 120
CERM Chapter 83: Construction and Job Site Safety 120

Systems, Management, and Professional

CERM Chapter 86: Project Management, Budgeting, and Scheduling. 125
CERM Chapter 87: Engineering Economic Analysis. 125

Index 129

How to Use This Book

This book (and others in the *Quick Reference* series) was developed to help you minimize problem-solving time on the PE exam. This *Quick Reference* is a consolidation of the most useful equations, tables, and figures in the *PE Civil Reference Manual*. Using *Quick Reference*, you will not need to wade through pages of descriptive text to find the formulas that remind you of your next solution step.

The idea is this: you study for the exam using (primarily) your *Reference Manual*, and you take the exam using (primarily) your *Quick Reference*.

This book follows the same order, has the same nomenclature, and uses the same chapter, section, equation, figure, and table numbers as the *PE Civil Reference Manual*. Once you become familiar with the sequencing of subjects in the *Reference Manual*, you will be at home with *Quick Reference*. Furthermore, you can always go to the corresponding chapter and section number in the *Reference Manual* for additional information about a subject.

Once you have studied and mastered the theory behind an exam topic, you are ready to tackle the task of solving practice problems. By providing the essential formulas to solve most problems, *Quick Reference* will help you save time solving problems right from the start. In fact, as you progress in your understanding of topics, you will find that you can rely more and more on *Quick Reference* for rapid retrieval of formulas—without needing to refer back to the *Reference Manual*.

When solving problems, you will find that there will be times when you need access to two kinds of information simultaneously: formulas and data, formulas and nomenclature, or formulas and theory. It is likely you have experienced the frustration of having to work problems while substituting a spare pencil, calculator, or shoelace for a page marker in a single book. With *Quick Reference*, you have a convenient way to keep the equations you need in front of you, even as you may be flip-flopping between formulas, theory, and data in your other references.

Quick Reference also provides a convenient place for neatnik engineers to add comments and reminders to equations, without having to mess up a primary reference. I expect you to recycle this book after the exam, so go ahead and write in it.

Once you start incorporating *Quick Reference* into your problem-solving routine, I predict you will not want to return to the one-book approach. *Quick Reference* will save you precious time—and that is how to use this book.

Codes and Standards

The information that was used to write and update this book was based on the exam specifications at the time of publication. However, as with engineering practice itself, the PE exam is not always based on the most current codes or cutting-edge technology. Similarly, codes, standards, and regulations adopted by state and local agencies often lag issuance by several years. It is likely that the codes that are most current, the codes that you use in practice, and the codes that are the basis of your exam will all be different.

PPI lists on its website the dates and editions of the codes, standards, and regulations on which NCEES has announced the PE exams are based. It is your responsibility to find out which codes are relevant to your exam.

CONSTRUCTION DESIGN STANDARDS

ACI 347: *Guide to Formwork for Concrete*, 2014. American Concrete Institute, Farmington Hills, MI (as an appendix of ACI SP-4, Eighth ed.).

ACI SP-4: *Formwork for Concrete*, Eighth ed., 2014. American Concrete Institute, Farmington Hills, MI.

ACI MNL-15: *Field Reference Manual*, 2016. American Concrete Institute, Farmington Hills, MI.

AISC: *Steel Construction Manual*, Fourteenth ed., 2011. American Institute of Steel Construction, Inc., Chicago, IL.

ASCE 37: *Design Loads on Structures During Construction*, 2014. American Society of Civil Engineers, Reston, VA.

CMWB: *Standard Practice for Bracing Masonry Walls Under Construction*, 2012. Council for Masonry Wall Bracing, Mason Contractors Association of America, Lombard, IL.

MUTCD-Pt 6: *Manual on Uniform Traffic Control Devices*—Part 6, Temporary Traffic Control, 2009. U.S. Department of Transportation, Federal Highway Administration, Washington, DC.

OSHA 1903, 1904, and 1926: *Construction Industry Regulations* (U.S. Federal version, January 2017). U.S. Department of Labor, Washington, DC.

GEOTECHNICAL DESIGN STANDARDS

ASCE/SEI7: *Minimum Design Loads for Buildings and Other Structures*, 2010. American Society of Civil Engineers, Reston, VA.

OSHA 1926: *Occupational Safety and Health Regulations for the Construction Industry* (U.S. Federal version), Subpart P, Excavations, Part 1926.651: Specific Excavation Requirements, and Subpart P, Excavations, Part 1926.652: Requirements for Protective Systems. U.S. Department of Labor, Washington, DC.

STRUCTURAL DESIGN STANDARDS

AASHTO LRFD: *AASHTO LRFD Bridge Design Specifications*, Seventh ed., 2016. American Association of State Highway and Transportation Officials, Washington, DC.

ACI 318: *Building Code Requirements for Structural Concrete*, 2014. American Concrete Institute, Farmington Hills, MI.

ACI 530/530.1[1]: *Building Code Requirements and Specification for Masonry Structures* (and companion commentaries), 2013. The Masonry Society, Boulder, CO; American Concrete Institute, Detroit, MI; and Structural Engineering Institute of the American Society of Civil Engineers, Reston, VA.

AISC: *Steel Construction Manual*, Fourteenth ed., 2011. American Institute of Steel Construction, Inc., Chicago, IL.

ASCE/SEI7: *Minimum Design Loads for Buildings and Other Structures*, third printing, 2010. American Society of Civil Engineers, Reston, VA.

AWS D1.1/D1.1M[2]: *Structural Welding Code—Steel*, Twenty-second ed., 2010. American Welding Society, Miami, FL.

AWS D1.2/D1.2M[3]: *Structural Welding Code—Aluminum*, Sixth ed., 2014. American Welding Society, Miami, FL.

---

[1]Only the Allowable Stress Design (ASD) method may be used on the structural depth exam, except that ACI 530 Sec. 9.3.5 (strength design) may be used for walls with out-of-plane loads.
[2]AWS D1.1, AWS D1.2, and AWS D1.4 are listed in the Codes, Standards, and Documents subsection of NCEES's Civil PE structural depth exam specifications.
[3]See Ftn. 2.

AWS D1.4/D1.4M[4]: *Structural Welding Code—Reinforcing Steel*, Seventh ed., 2011. American Welding Society, Miami, FL.

IBC: *2015 International Building Code* (without supplements). International Code Council, Inc., Falls Church, VA.

NDS[5]: *National Design Specification for Wood Construction ASD/LRFD*, 2015 ed., and *National Design Specification Supplement, Design Values for Wood Construction*, 2015 ed. American Wood Council, Washington, DC.

OSHA 1910: *Occupational Safety and Health Standards* (U.S. Federal version)Subpart A, General, 1910.1–1910.9, with Appendix A to 1910.7; Subpart D, Walking-Working Surfaces, 1910.21–1910.30; Subpart F, Powered Platforms, Manlifts, and Vehicle-Mounted Work Platforms, 1910.66–1910.68, with Appendix A–Appendix D to 1910.66. U.S. Department of Labor, Washington, DC.

OSHA 1926: *Occupational Safety and Health Regulations for the Construction Industry* (U.S. Federal version) Subpart E, Personal Protective and Life Saving Equipment, 1926.95–1926.107; Subpart M, Fall Protection, 1926.500–1926.503, App. A–E; Subpart Q, Concrete and Masonry Construction, 1926.700–1926.706, with App. A; and Subpart R, Steel Erection, 1926.750–1926.761, with App. A–H. U.S. Department of Labor, Washington, DC.

PCI: *PCI Design Handbook: Precast and Prestressed Concrete*, Seventh ed., 2010. Precast/Prestressed Concrete Institute, Chicago, IL.

TRANSPORTATION DESIGN STANDARDS

AASHTO *GDPS*: *AASHTO Guide for Design of Pavement Structures* (GDPS-4-M), 1993, and 1998 supplement. American Association of State Highway and Transportation Officials, Washington, DC.

AASHTO *Green Book*: *A Policy on Geometric Design of Highways and Streets*, Sixth ed., 2011 (including November 2013 errata). American Association of State Highway and Transportation Officials, Washington, DC.

AASHTO: *Guide for the Planning, Design, and Operation of Pedestrian Facilities*, First ed., 2004. American Association of State Highway and Transportation Officials, Washington, DC.

HSM: *Highway Safety Manual*, First ed., 2010 vols. 1-3 (including September 2010, Febrary 2012, and March 2016 errata). American Association of State Highway and Transportation Officials, Washington, DC.

AASHTO *MEPDG*: *Mechanistic-Empirical Pavement Design Guide: A Manual of Practice*, Interim ed., 2008. American Association of State Highway and Transportation Officials, Washington, DC.

AASHTO: *Roadside Design Guide*, Fourth ed., 2011 (including February 2012 and July 2015 errata). American Association of State Highway and Transportation Officials, Washington, DC.

AI: *The Asphalt Handbook* (MS-4), Seventh ed., 2007. Asphalt Institute, Lexington, KY.

FHWA: *Hydraulic Design of Highway Culverts*, Hydraulic Design Series no. 5, Publication no. FHWA-HIF-12-026, Third ed., 2012. U.S. Department of Transportation, Federal Highway Administration, Washington, DC.

HCM: *Highway Capacity Manual*, 2010 ed., vols 1-3, including Approved HCM 2010 Corrections and Clarifications (as of January 2014); Approved HCM 2010 Interpretations (as of January 2014); Replacement HCM 2010 Volume 1–3 pages (April 2014); Replacement HCM 2010 Volume 1–3 pages (January 12–February 13); Replacement HCM 2010 Volume 1–3 pages (March 2013). Transportation Research Board, National Research Council, Washington, DC.

MUTCD: *Manual on Uniform Traffic Control Devices*, 2009 (including Revisions 1 and 2, May 2012). U.S. Department of Transportation, Federal Highway Administration, Washington, DC.

PCA: *Design and Control of Concrete Mixtures*, Sixteenth ed., 2016. Portland Cement Association, Skokie, IL.

---

[4]See Ftn. 2.
[5]Only the ASD method may be used for wood design on the structural depth exam.

For Instant Recall

Fundamental and Physical Constants

quantity	symbol	English		SI
Density				
air [STP] [32°F (0°C)]		0.0805 lbm/ft^3		1.29 kg/m^3
air [70°F (20°C), 1 atm]		0.0749 lbm/ft^3		1.20 kg/m^3
earth [mean]		345 lbm/ft^3		5520 kg/m^3
mercury		849 lbm/ft^3		1.360×10^4 kg/m^3
seawater		64.0 lbm/ft^3		1025 kg/m^3
water [mean]		62.4 lbm/ft^3		1000 kg/m^3
Specific Gravity				
mercury			13.6	
water			1.0	
Gravitational Acceleration				
earth [mean]	g	32.174 (32.2) ft/sec^2		9.8067 (9.81) m/s^2
moon [mean]		5.47 ft/sec^2		1.67 m/s^2
Pressure, atmospheric		14.696 (14.7) lbf/in^2		1.0133×10^5 Pa
Temperature, standard		32°F (492°R)		0°C (273K)
Fundamental Constants				
Avogadro's number	N_A			6.022×10^{23} mol^{-1}
gravitational constant	g_c	32.174 lbm-ft/lbf-sec^2		
gravitational constant	G	3.430×10^{-8} ft^4/lbf-sec^4		6.674×10^{-11} N·m^2/kg^2
specific gas constant, air	R	53.35 ft-lbf/lbm-°R		287.03 J/kg·K
specific gas constant, methane	R_{CH_4}	96.32 ft-lbf/lbm-°R		518.3 J/kg·K
triple point, water		32.02°F, 0.0888 psia		0.01109°C, 0.6123 kPa
universal gas constant	R^*	1545.35 ft-lbf/lbmol-°R		8314.47 J/kmol·K
	R^*	1.986 Btu/lbmol-°R		0.08206 atm·L/mol·K
Molecular Weight				
air			29	
carbon			12	
carbon dioxide			44	
helium			4	
hydrogen			2	
methane			16	
nitrogen			28	
oxygen			32	
Steel				
modulus of elasticity		2.9×10^7 psi		
modulus of shear		1.2×10^7 psi		
Poisson's ratio			0.3	

Temperature Conversions

$$°F = 32° + \frac{9}{5}°C$$
$$°C = \frac{5}{9}(°F - 32°)$$
$$°R = °F + 460°$$
$$K = °C + 273°$$
$$\Delta°R = \frac{9}{5}\Delta K$$
$$\Delta K = \frac{5}{9}\Delta°R$$

SI Prefixes

symbol	prefix	value
a	atto	10^{-18}
f	femto	10^{-15}
p	pico	10^{-12}
n	nano	10^{-9}
μ	micro	10^{-6}
m	milli	10^{-3}
c	centi	10^{-2}
d	deci	10^{-1}
da	deka	10
h	hecto	10^{2}
k	kilo	10^{3}
M	mega	10^{6}
G	giga	10^{9}
T	tera	10^{12}
P	peta	10^{15}
E	exa	10^{18}

Equivalent Units of Derived and Common SI Units

symbol	equivalent units						
A	C/s	W/V	V/Ω	$J/s{\cdot}V$	$N/T{\cdot}m$	Wb/H	$kg{\cdot}m^2/V{\cdot}s^2$
C	$A{\cdot}s$	J/V	$N{\cdot}m/V$	$V{\cdot}F$			
F	C/V	C^2/J	$C^2/N{\cdot}m$	$A^2{\cdot}s^4/kg{\cdot}m^2$	$kg{\cdot}m^2/V{\cdot}s^2$		
F/m	$C/V{\cdot}m$	$C^2/J{\cdot}m$	$C^2/N{\cdot}m^2$	$A^2{\cdot}s^4/kg{\cdot}m^3$			
H	Wb/A	$V{\cdot}s/A$	$T{\cdot}m^2/A$	$kg{\cdot}m^2/A^2{\cdot}s^2$	$\Omega{\cdot}s$		
Hz	1/s						
J	$N{\cdot}m$	$V{\cdot}C$	$W{\cdot}s$	C^2/F	$kg{\cdot}m^2/s^2$		
m^2/s^2	$V{\cdot}C/kg$						
N	J/m	$V{\cdot}C/m$	$kg{\cdot}m/s^2$				
N/A^2	$Wb/N{\cdot}m^2$	$kg{\cdot}m/A^2{\cdot}s^2$					
Pa	N/m^2	$kg/m{\cdot}s^2$					
Ω	V/A	$kg{\cdot}m^2/A^2{\cdot}s^3$					
S	A/V	$A^2{\cdot}s^3/kg{\cdot}m^2$					
T	Wb/m^2	$N/A{\cdot}m$	$N{\cdot}s/C{\cdot}m$	$kg/A{\cdot}s^2$			
V	W/A	C/F	J/C	$kg{\cdot}m^2/A{\cdot}s^3$			
V/m	N/C	$J/m{\cdot}C$	$kg{\cdot}m/s^2{\cdot}C$	$kg{\cdot}m/A{\cdot}s^3$			
W	J/s	$V{\cdot}A$	$kg{\cdot}m^2/s^3$	$N{\cdot}m/s$			
Wb	$V{\cdot}s$	$H{\cdot}A$	$T{\cdot}m^2$	$kg{\cdot}m^2/A{\cdot}s^2$			

Quick
ENGINEERING CONVERSIONS
Unless noted otherwise, atmospheres are standard; Btus are IT (international table); calories are gram-calories;
gallons are U.S. liquid; miles are statute; pounds-mass are avoirdupois; chains are surveyors'.

multiply	by	to obtain
ac	10.0	chain^2
ac	43,560	ft^2
ac	0.40469	hectare
ac	4046.87	m^2
ac	1/640	mi^2
ac-ft	43,560	ft^3
ac-ft	0.01875	mi^2-in
ac-ft	1233.5	m^3
ac-ft	325,851	gal
ac-ft/day	0.50416	ft^3/s
ac-ft/mi^2	0.01875	in (runoff)
ac-in	102,790	L
ac-in/hr	1.0083	ft^3/s
angstrom	1.0×10^{-10}	m
atm	1.01325	bar
atm	76.0	cm Hg
atm	33.90	ft water
atm	29.921	in Hg
atm	14.696	lbf/in^2
atm	101.33	kPa
atm	1.0133×10^5	Pa
bar	0.9869	atm
bar	1.0×10^5	Pa
Btu	778.26	ft-lbf
Btu	1055.056	J
Btu	2.928×10^{-4}	kW-hr
Btu	1.0×10^{-5}	therm
Btu/hr	0.21611	ft-lbf/s
Btu/hr	3.929×10^{-4}	hp
Btu/hr	0.29307	W
Btu/lbm	2.3260	kJ/kg
Btu/lbm-°R	4.1868	kJ/kg·K
Btu/s	778.26	ft-lbf/s
Btu/s	46,680	ft-lbf/min
Btu/s	1.4148	hp
Btu/s	1.0545	kW
cal	3.968×10^{-3}	Btu
cal	4.1868	J
chain	66.0	ft
chain	1/80	mi (statute)
chain	4.0	rod
chain	22	yd
chain^2	1/10	ac
cm	0.03281	ft
cm	0.39370	in
cm/s	1.9686	ft/min
cm^2	0.1550	in^2
cm^3	2.6417×10^{-4}	gal
cm^3	0.0010	L
day (mean solar)	86,400	s
day (sidereal)	86,164.09	s
darcy	1.0623×10^{-11}	ft^2
degree (angular)	1/0.9	grad
degree (angular)	$2\pi/360$	radian
degree (angular)	17.778	mil
dyne	1.0×10^{-5}	N
eV	1.6022×10^{-19}	J
ft	1/66	chain
ft	0.3048	m
ft	1/5280	mi
ft	0.06060	rod
ft of water	0.43328	lbf/in^2
ft-kips	1.358	kN·m
ft-lbf (energy)	1.2851×10^{-3}	Btu
ft-lbf (energy)	1.3558	J
ft-lbf (energy)	3.766×10^{-7}	kW·h
ft-lbf (torque)	1.3558	N·m
ft-lbf/min	1/46,680	Btu/s
ft-lbf/min	1/33,000	hp
ft-lbf/min	2.2598×10^{-5}	kW
ft-lbf/s	1.2849×10^{-3}	Btu/s
ft-lbf/s	1/550	hp

multiply	by	to obtain
ft-lbf/s	1.3558×10^{-3}	kW
ft/min	0.50798	cm/s
ft/s	0.68180	mi/hr
ft^2	1/43,560	ac
ft^2	9.4135×10^{10}	darcy
ft^2	0.09290	m^2
ft^3	1/43,560	ac-ft
ft^3	1728.0	in^3
ft^3	7.4805	gal
ft^3	28.320	L
ft^3	0.03704	yd^3
ft^3/day	0.005195	gal/min
ft^3/mi^2	4.3×10^{-7}	in runoff
ft^3/min	10,772	gal/day
ft^3/s	1.9835	ac-ft/day
ft^3/s	0.99177	ac-in/hr
ft^3/s	646,317	gal/day
ft^3/s	448.83	gal/min
ft^3/s	0.03719	mi^2-in/day
ft^3/s	0.64632	MGD
g	0.03527	oz
g	0.002205	lbm
g/cm^3	1000.0	kg/m^3
g/cm^3	62.428	lbm/ft^3
gal	1/325,851	ac-ft
gal	3785.4	cm^3
gal	0.13368	ft^3
gal (Imperial)	1.2	gal (U.S.)
gal (U.S.)	0.8327	gal (Imperial)
gal	3.7854	L
gal	3.7854×10^{-3}	m^3
gal	1.0×10^{-6}	MG
gal	0.00495	yd^3
gal (of water)	8.34	lbm (of water)
gal/ac-day	0.04356	MG/ft^2-day
gal/day	9.283×10^{-5}	ft^3/min
gal/day	1.5472×10^{-6}	ft^3/s
gal/day	1/1440	gal/min
gal/day	1.0×10^{-6}	MG/day
gal/day-ft	0.01242	m^3/m·day
gal/day-ft^2	0.04075	m^3/m^2·day
gal/day-ft^2	1.0	Meinzer unit
gal/day-ft^2	0.04355	MGD/ac (mgad)
gal/min	0.002228	ft^3/s
gal/min	192.50	ft^3/day
gal/min	1440	gal/day
gal/min	0.06309	L/s
grad	0.90	degrees (angular)
grain	1.4286×10^{-4}	lbm
grain/gal	142.86	lbm/MG
grain/gal	17.118	ppm
grain/gal	17.118	mg/L
hectare	2.4711	ac
hectare	10,000	m^2
hp	550.0	ft-lbf/s
hp	33,000	ft-lbf/min
hp	0.7457	kW
hp	2545	Btu/hr
hp	0.70678	Btu/s
hp-hr	2545.2	Btu
in	2.540	cm
in	0.02540	m
in	25.40	mm
in Hg	0.4910	lbf/in^2
in Hg	70.704	lbf/ft^2
in Hg	13.60	in water
in runoff	53.30	ac-ft/mi^2
in runoff	2.3230×10^6	ft^3/mi^2
in water	5.1990	lbf/ft^2
in water	0.0361	lbf/in^2
in water	0.07353	in Hg
in-lbf	0.11298	N·m

multiply	by	to obtain
in/ft	1/0.012	mm/m
in^2	6.4516	cm^2
in^3	1/1728	ft^3
J	9.4778×10^{-4}	Btu
J	6.2415×10^{18}	eV
J	0.73756	ft-lbf
J	1.0	N·m
J/s	1.0	W
kg	2.2046	lbm
kg/m^3	0.06243	lbm/ft^3
kip	4.4480	kN
kip	1000.0	lbf
kip	4448.0	N
kip/ft	14.594	kN/m
kip/ft^2	47.880	kPa
kJ	0.94778	Btu
kJ	737.56	ft-lbf
kJ/kg	0.42992	Btu/lbm
kJ/kg·K	0.23885	Btu/lbm-°R
km	3280.8	ft
km	0.62138	mi
km/hr	0.62138	mi/hr
kN	0.2248	kips
kN·m	0.73757	ft-kips
kN/m	0.06852	kips/ft
kPa	9.8692×10^{-3}	atm
kPa	0.14504	lbf/in^2
kPa	1000.0	Pa
kPa	0.02089	kips/ft^2
kPa	0.01	bar
ksi	6.8948×10^{6}	Pa
ksi	6894.8	kPa
kW	737.56	ft-lbf/s
kW	44,250	ft-lbf/min
kW	1.3410	hp
kW	3413.0	Btu/hr
kW	0.9483	Btu/s
kW-hr	3413.0	Btu
kW-hr	3.60×10^{6}	J
L	1/102,790	ac-in
L	1000.0	cm^3
L	0.03531	ft^3
L	0.26417	gal
L	61.024	in^3
L	0.0010	m^3
L/s	2.1189	ft^3/min
L/s	15.850	gal/min
lbf	0.001	kips
lbf	4.4482	N
lbf/ft^2	0.01414	in Hg
lbf/ft^2	0.19234	in water
lbf/ft^2	0.00694	lbf/in^2
lbf/ft^2	47.880	Pa
lbf/ft^2	5.0×10^{-4}	tons/ft^2
lbf/in^2	0.06805	atm
lbf/in^2	144.0	lbf/ft^2
lbf/in^2	2.308	ft water
lbf/in^2	27.70	in water
lbf/in^2	2.0370	in Hg
lbf/in^2	6894.8	Pa
lbf/in^2	0.00050	tons/in^2
lbf/in^2	0.0720	tons/ft^2
lbm	7000.0	grains
lbm	453.59	g
lbm	0.45359	kg
lbm	4.5359×10^{5}	mg
lbm	5.0×10^{-4}	tons (mass)
lbm (of water)	0.12	gal (of water)
lbm/ac-ft-day	0.02296	lbm/1000 ft^3-day
lbm/ft^3	0.016018	g/cm^3
lbm/ft^3	16.018	kg/m^3
lbm/1000 ft^3-day	43.560	lbm/ac-ft-day
lbm/1000 ft^3-day	133.68	lbm/MG-day
lbm/MG	0.0070	grains/gal
lbm/MG	0.11983	mg/L
lbm/MG-day	0.00748	lbm/1000 ft^3-day
leagues	4428.0	m

multiply	by	to obtain
m	3.2808	ft
m	39.370	in
m	2.2583×10^{-4}	leagues
m	1.0936	yd
m/s	196.85	ft/min
m^2	2.4711×10^{-4}	ac
m^2	10.764	ft^2
m^2	1/10,000	hectare
mi^2-in	53.3	ac-ft
mi^2-in/day	26.89	ft^3/s
m^3	8.1071×10^{-4}	ac-ft
m^3/m·day	80.5196	gal/day-ft
m^3/m^2·day	24.542	gal/day-ft^2
Meinzer unit	1.0	gal/day-ft^2
mg	2.2046×10^{-6}	lbm
mg/L	1.0	ppm
mg/L	0.05842	grains/gal
mg/L	8.3454	lbm/MG
MG	1.0×10^{6}	gal
MG/ac-day	22.968	gal/ft^2-day
MGD	1.5472	ft^3/sec
MGD	1×10^{6}	gal/day
MGD/ac (mgad)	22.957	gal/day-ft^2
mi	5280.0	ft
mi	80.0	chains
mi	1.6093	km
mi (statute)	0.86839	miles (nautical)
mi	320.0	rods
mi/hr	1.4667	ft/s
mi^2	640.0	acres
micron	1.0×10^{-6}	m
micron	0.001	mm
mil (angular)	0.05625	degrees
mil (angular)	3.375	min
min (angular)	0.29630	mils
min (angular)	2.90888×10^{-4}	radians
min (time, mean solar)	60	s
mm	1/25.4	in
mm	1000.0	microns
mm/m	0.012	in/ft
MPa	1.0×10^{6}	Pa
N	0.22481	lbf
N	1.0×10^{5}	dynes
N·m	0.73756	ft-lbf
N·m	8.8511	in-lbf
N·m	1.0	J
N/m^2	1.0	Pa
oz	28.353	g
Pa	0.001	kPa
Pa	1.4504×10^{-7}	ksi
Pa	1.4504×10^{-4}	lbf/in^2
Pa	0.02089	lbf/ft^2
Pa	1.0×10^{-6}	MPa
Pa	1.0	N/m^2
ppm	0.05842	grains/gal
radian	$180/\pi$	degrees (angular)
radian	3437.7	min (angular)
rod	0.250	chain
rod	16.50	ft
rod	1/320	mi
s (time)	1/86,400	day (mean solar)
s (time)	1.1605×10^{-5}	day (sidereal)
s (time)	1/60	min
therm	1.0×10^{5}	Btu
ton (force)	2000.0	lbf
ton (mass)	2000.0	lbm
ton/ft^2	2000.0	lbf/ft^2
ton/ft^2	13.889	lbf/in^2
W	3.413	Btu/hr
W	0.73756	ft-lbf/s
W	1.3410×10^{-3}	hp
W	1.0	J/s
yd	1/22	chain
yd	0.91440	m
yd^3	27	ft^3
yd^3	201.97	gal

Mensuration of Two-Dimensional Areas

Nomenclature

A total surface area
b base
c chord length
d distance
h height
L length
p perimeter
r radius
s side (edge) length, arc length
θ vertex angle, in radians
ϕ central angle, in radians

Circular Sector

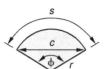

$$A = \tfrac{1}{2}\phi r^2 = \tfrac{1}{2}sr$$

$$\phi = \frac{s}{r}$$

$$s = r\phi$$

$$c = 2r\sin\left(\frac{\phi}{2}\right)$$

Triangle

equilateral   right   oblique

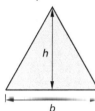 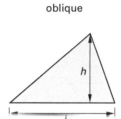

$$A = \tfrac{1}{2}bh = \frac{\sqrt{3}}{4}b^2 \qquad A = \tfrac{1}{2}bh \qquad\qquad A = \tfrac{1}{2}bh$$

$$h = \frac{\sqrt{3}}{2}b \qquad\qquad H^2 = b^2 + h^2$$

Parabola

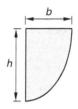

$$A = \tfrac{2}{3}bh$$

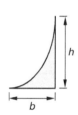

$$A = \tfrac{1}{3}bh$$

Circle

$$p = 2\pi r$$

$$A = \pi r^2 = \frac{p^2}{4\pi}$$

Circular Segment

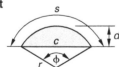

$$A = \tfrac{1}{2}r^2(\phi - \sin\phi)$$

$$\phi = \frac{s}{r} = 2\left(\arccos\frac{r-d}{r}\right)$$

$$c = 2r\sin\left(\frac{\phi}{2}\right)$$

Ellipse

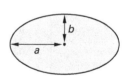

$$A = \pi a b$$

$$p \approx 2\pi\sqrt{\tfrac{1}{2}(a^2 + b^2)} \qquad \begin{bmatrix} \text{Euler's} \\ \text{upper bound} \end{bmatrix}$$

Trapezoid

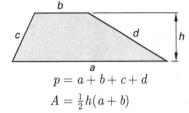

$$p = a + b + c + d$$
$$A = \tfrac{1}{2}h(a + b)$$

If $c = d$, the trapezoid is isosceles.

Parallelogram

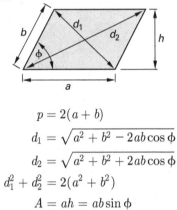

$$p = 2(a + b)$$
$$d_1 = \sqrt{a^2 + b^2 - 2ab\cos\phi}$$
$$d_2 = \sqrt{a^2 + b^2 + 2ab\cos\phi}$$
$$d_1^2 + d_2^2 = 2(a^2 + b^2)$$
$$A = ah = ab\sin\phi$$

If $a = b$, the parallelogram is a rhombus.

Regular Polygon (*n* equal sides)

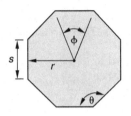

$$\phi = \frac{2\pi}{n}$$
$$\theta = \frac{\pi(n-2)}{n} = \pi - \phi$$
$$p = ns$$
$$s = 2r\tan\frac{\theta}{2}$$
$$A = \tfrac{1}{2}nsr$$

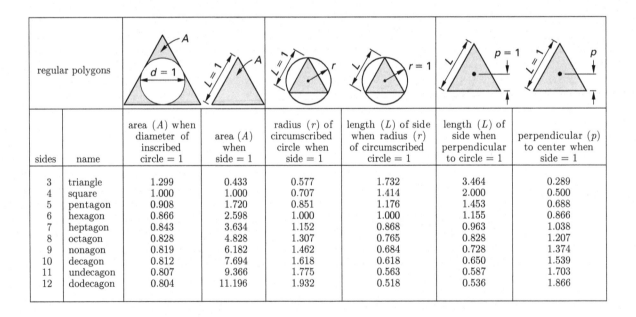

sides	name	area (A) when diameter of inscribed circle = 1	area (A) when side = 1	radius (r) of circumscribed circle when side = 1	length (L) of side when radius (r) of circumscribed circle = 1	length (L) of side when perpendicular to circle = 1	perpendicular (p) to center when side = 1
3	triangle	1.299	0.433	0.577	1.732	3.464	0.289
4	square	1.000	1.000	0.707	1.414	2.000	0.500
5	pentagon	0.908	1.720	0.851	1.176	1.453	0.688
6	hexagon	0.866	2.598	1.000	1.000	1.155	0.866
7	heptagon	0.843	3.634	1.152	0.868	0.963	1.038
8	octagon	0.828	4.828	1.307	0.765	0.828	1.207
9	nonagon	0.819	6.182	1.462	0.684	0.728	1.374
10	decagon	0.812	7.694	1.618	0.618	0.650	1.539
11	undecagon	0.807	9.366	1.775	0.563	0.587	1.703
12	dodecagon	0.804	11.196	1.932	0.518	0.536	1.866

Mensuration of Three-Dimensional Volumes

Nomenclature
A surface area
b base
h height
r radius
R radius
s side (edge) length
V internal volume

Sphere
$$V = \tfrac{4}{3}\pi r^3 = \tfrac{4}{3}\pi \left(\frac{d}{2}\right)^3 = \tfrac{1}{6}\pi d^3$$
$$A = 4\pi r^2$$

Right Circular Cone (excluding base area)
$$V = \tfrac{1}{3}\pi r^2 h = \tfrac{1}{3}\pi \left(\frac{d}{2}\right)^2 h = \tfrac{1}{12}\pi d^2 h$$
$$A = \pi r\sqrt{r^2 + h^2}$$

Right Circular Cylinder (excluding end areas)
$$V = \pi r^2 h$$
$$A = 2\pi r h$$

Spherical Segment (spherical cap)

Surface area of a spherical segment of radius r cut out by an angle θ_0 rotated from the center about a radius, r, is
$$A = 2\pi r^2 (1 - \cos\theta_0)$$
$$\omega = \frac{A}{r^2} = 2\pi(1 - \cos\theta_0)$$

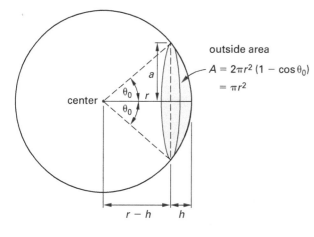

outside area
$A = 2\pi r^2 (1 - \cos\theta_0)$
$= \pi r^2$

$$V_{\mathrm{cap}} = \tfrac{1}{6}\pi h(3a^2 + h^2)$$
$$= \tfrac{1}{3}\pi h^2 (3r - h)$$
$$a = \sqrt{h(2r - h)}$$

Paraboloid of Revolution

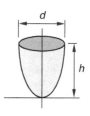

$$V = \tfrac{1}{8}\pi d^2 h$$

Torus

$$A = 4\pi^2 R r$$
$$V = 2\pi^2 R r^2$$

Regular Polyhedra (identical faces)

name	number of faces	form of faces	total surface area	volume
tetrahedron	4	equilateral triangle	$1.7321\,s^2$	$0.1179\,s^3$
cube	6	square	$6.0000\,s^2$	$1.0000\,s^3$
octahedron	8	equilateral triangle	$3.4641\,s^2$	$0.4714\,s^3$
dodecahedron	12	regular pentagon	$20.6457\,s^2$	$7.6631\,s^3$
icosahedron	20	equilateral triangle	$8.6603\,s^2$	$2.1817\,s^3$

The radius of a sphere inscribed within a regular polyhedron is

$$r = \frac{3\,V_{\mathrm{polyhedron}}}{A_{\mathrm{polyhedron}}}$$

Background and Support

CERM Chapter 3
Algebra

9. ROOTS OF QUADRATIC EQUATIONS

$$x_1, x_2 = \frac{-b \pm \sqrt{b^2 - 4ac}}{2a}$$ *3.17*

13. RULES FOR EXPONENTS AND RADICALS

$$(ab)^n = a^n b^n$$ *3.24*

$$b^{m/n} = \sqrt[n]{b^m} = \left(\sqrt[n]{b}\right)^m$$ *3.25*

$$\left(b^n\right)^m = b^{nm}$$ *3.26*

$$b^m b^n = b^{m+n}$$ *3.27*

15. LOGARITHM IDENTITIES

$$\log_b b = 1$$ *3.34*

$$\log_b 1 = 0$$ *3.35*

$$\log_b b^n = n$$ *3.36*

$$\log x^a = a \log x$$ *3.37*

$$\log xy = \log x + \log y$$ *3.41*

$$\log \frac{x}{y} = \log x - \log y$$ *3.42*

$$\log_a x = \log_b x \log_a b$$ *3.43*

$$\ln x = \ln 10 \log_{10} x \approx 2.3026 \log_{10} x$$ *3.44*

$$\log_{10} x = \log_{10} \ln x \, e \approx 0.4343 \ln x$$ *3.45*

CERM Chapter 4
Linear Algebra

5. DETERMINANTS

$$\mathbf{A} = \begin{bmatrix} a & b \\ c & d \end{bmatrix}$$

$$|\mathbf{A}| = \begin{vmatrix} a & b \\ c & d \end{vmatrix} = ad - bc$$ *4.3*

$$\mathbf{A} = \begin{bmatrix} a & b & c \\ d & e & f \\ g & h & i \end{bmatrix}$$

$$|\mathbf{A}| = a \begin{vmatrix} e & f \\ h & i \end{vmatrix} - d \begin{vmatrix} b & c \\ h & i \end{vmatrix} + g \begin{vmatrix} b & c \\ e & f \end{vmatrix}$$ *4.6*

CERM Chapter 5
Vectors

2. VECTORS IN *n*-SPACE

$$|\mathbf{V}| = \sqrt{(x_2 - x_1)^2 + (y_2 - y_1)^2}$$ *5.1*

$$\phi = \arctan \frac{y_2 - y_1}{x_2 - x_1}$$ *5.2*

Figure 5.1 *Vector in Two-Dimensional Space*

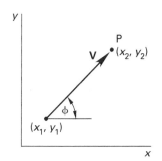

3. UNIT VECTORS

$$\mathbf{V} = |\mathbf{V}|\mathbf{a} = V_x\mathbf{i} + V_y\mathbf{j} + V_z\mathbf{k} \qquad 5.8$$

Figure 5.3 *Cartesian Unit Vectors*

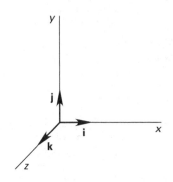

8. VECTOR DOT PRODUCT

$$\mathbf{V}_1 \cdot \mathbf{V}_2 = |\mathbf{V}_1|\,|\mathbf{V}_2|\cos\phi$$
$$= V_{1x}V_{2x} + V_{1y}V_{2y} + V_{1z}V_{2z} \qquad 5.21$$

Figure 5.5 *Vector Dot Product*

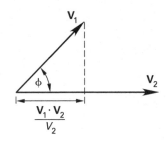

9. VECTOR CROSS PRODUCT

$$|\mathbf{V}_1 \times \mathbf{V}_2| = |\mathbf{V}_1||\mathbf{V}_2|\sin\phi \qquad 5.32$$

Figure 5.6 *Vector Cross Product*

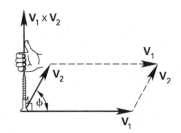

CERM Chapter 6
Trigonometry

1. DEGREES AND RADIANS

multiply	by	to obtain
radians	$\dfrac{180}{\pi}$	degrees
degrees	$\dfrac{\pi}{180}$	radians

4. RIGHT TRIANGLES

$$x^2 + y^2 = r^2 \qquad 6.1$$

5. CIRCULAR TRANSCENDENTAL FUNCTIONS

$$\text{sine: } \sin\theta = \frac{y}{r} = \frac{\text{opposite}}{\text{hypotenuse}} \qquad 6.2$$

$$\text{cosine: } \cos\theta = \frac{x}{r} = \frac{\text{adjacent}}{\text{hypotenuse}} \qquad 6.3$$

$$\text{tangent: } \tan\theta = \frac{y}{x} = \frac{\text{opposite}}{\text{adjacent}} \qquad 6.4$$

$$\text{cotangent: } \cot\theta = \frac{x}{y} = \frac{\text{adjacent}}{\text{opposite}} \qquad 6.5$$

$$\text{secant: } \sec\theta = \frac{r}{x} = \frac{\text{hypotenuse}}{\text{adjacent}} \qquad 6.6$$

$$\text{cosecant: } \csc\theta = \frac{r}{y} = \frac{\text{hypotenuse}}{\text{opposite}} \qquad 6.7$$

$$\cot\theta = \frac{1}{\tan\theta} \qquad 6.8$$

$$\sec\theta = \frac{1}{\cos\theta} \qquad 6.9$$

$$\csc\theta = \frac{1}{\sin\theta} \qquad 6.10$$

Figure 6.6 *Trigonometric Functions in a Unit Circle*

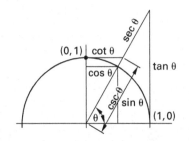

6. SMALL ANGLE APPROXIMATIONS

$$\sin\theta \approx \tan\theta \approx \theta\big|_{\theta < 10° \ (0.175 \ \text{rad})} \qquad 6.11$$

$$\cos\theta \approx 1\big|_{\theta < 5° \ (0.0873 \ \text{rad})} \qquad 6.12$$

10. TRIGONOMETRIC IDENTITIES

$$\sin^2\theta + \cos^2\theta = 1 \qquad 6.14$$

$$1 + \tan^2\theta = \sec^2\theta \qquad 6.15$$

$$1 + \cot^2\theta = \csc^2\theta \qquad 6.16$$

- *double-angle formulas*

$$\sin 2\theta = 2\sin\theta\cos\theta = \frac{2\tan\theta}{1+\tan^2\theta} \qquad 6.17$$

$$\cos 2\theta = \cos^2\theta - \sin^2\theta = 1 - 2\sin^2\theta$$
$$= 2\cos^2\theta - 1 = \frac{1-\tan^2\theta}{1+\tan^2\theta} \qquad 6.18$$

$$\tan 2\theta = \frac{2\tan\theta}{1-\tan^2\theta} \qquad 6.19$$

$$\cot 2\theta = \frac{\cot^2\theta - 1}{2\cot\theta} \qquad 6.20$$

- *two-angle formulas*

$$\sin(\theta \pm \phi) = \sin\theta\cos\phi \pm \cos\theta\sin\phi \qquad 6.21$$

$$\cos(\theta \pm \phi) = \cos\theta\cos\phi \mp \sin\theta\sin\phi \qquad 6.22$$

$$\tan(\theta \pm \phi) = \frac{\tan\theta \pm \tan\phi}{1 \mp \tan\theta\tan\phi} \qquad 6.23$$

$$\cot(\theta \pm \phi) = \frac{\cot\phi\cot\theta \mp 1}{\cot\phi \pm \cot\theta} \qquad 6.24$$

- *half-angle formulas* ($\theta < 180°$)

$$\sin\frac{\theta}{2} = \sqrt{\frac{1-\cos\theta}{2}} \qquad 6.25$$

$$\cos\frac{\theta}{2} = \sqrt{\frac{1+\cos\theta}{2}} \qquad 6.26$$

$$\tan\frac{\theta}{2} = \sqrt{\frac{1-\cos\theta}{1+\cos\theta}} = \frac{\sin\theta}{1+\cos\theta} = \frac{1-\cos\theta}{\sin\theta} \qquad 6.27$$

- *miscellaneous formulas* ($\theta < 90°$)

$$\sin\theta = 2\sin\frac{\theta}{2}\cos\frac{\theta}{2} \qquad 6.28$$

$$\sin\theta = \sqrt{\frac{1-\cos 2\theta}{2}} \qquad 6.29$$

$$\cos\theta = \cos^2\frac{\theta}{2} - \sin^2\frac{\theta}{2} \qquad 6.30$$

$$\cos\theta = \sqrt{\frac{1+\cos 2\theta}{2}} \qquad 6.31$$

$$\tan\theta = \frac{2\tan\frac{\theta}{2}}{1-\tan^2\frac{\theta}{2}} = \frac{2\sin\frac{\theta}{2}\cos\frac{\theta}{2}}{\cos^2\frac{\theta}{2} - \sin^2\frac{\theta}{2}} \qquad 6.32$$

$$\tan\theta = \sqrt{\frac{1-\cos 2\theta}{1+\cos 2\theta}} = \frac{\sin 2\theta}{1+\cos 2\theta} = \frac{1-\cos 2\theta}{\sin 2\theta} \qquad 6.33$$

$$\cot\theta = \frac{\cot^2\frac{\theta}{2} - 1}{2\cot\frac{\theta}{2}}$$
$$= \frac{\cos^2\frac{\theta}{2} - \sin^2\frac{\theta}{2}}{2\sin\frac{\theta}{2}\cos\frac{\theta}{2}} \qquad 6.34$$

$$\cot\theta = \sqrt{\frac{1+\cos 2\theta}{1-\cos 2\theta}} = \frac{1+\cos 2\theta}{\sin 2\theta} = \frac{\sin 2\theta}{1-\cos 2\theta} \qquad 6.35$$

12. ARCHAIC FUNCTIONS

$$\text{vers}\,\theta = 1 - \cos\theta = 2\sin^2\frac{\theta}{2} \qquad 6.36$$

$$\text{covers}\,\theta = 1 - \sin\theta \qquad 6.37$$

$$\text{havers}\,\theta = \frac{1-\cos\theta}{2} = \sin^2\frac{\theta}{2} \qquad 6.38$$

$$\text{exsec}\,\theta = \sec\theta - 1 \qquad 6.39$$

15. GENERAL TRIANGLES

$$\frac{\sin A}{a} = \frac{\sin B}{b} = \frac{\sin C}{c} \qquad 6.61$$

$$a^2 = b^2 + c^2 - 2bc\cos A \qquad 6.62$$

Figure 6.9 General Triangle

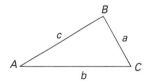

CERM Chapter 7
Analytic Geometry

2. AREAS WITH IRREGULAR BOUNDARIES

- *trapezoidal rule*

$$A = \frac{d}{2}\left(h_0 + h_n + 2\sum_{i=1}^{n-1} h_i\right) \qquad 7.1$$

- *Simpson's rule*

$$A = \frac{d}{3}\left(h_0 + h_n + 4\sum_{\substack{i \text{ odd} \\ i=1}}^{n-1} h_i + 2\sum_{\substack{i \text{ even} \\ i=2}}^{n-2} h_i\right) \qquad 7.2$$

Figure 7.1 *Irregular Areas*

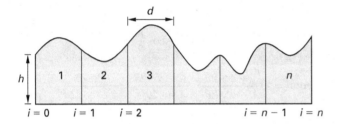

10. STRAIGHT LINES

- *general form*

$$Ax + By + C = 0 \qquad 7.8$$

$$A = -mB \qquad 7.9$$

$$B = \frac{-C}{b} \qquad 7.10$$

$$C = -aA = -bB \qquad 7.11$$

- *slope-intercept form*

$$y = mx + b \qquad 7.12$$

$$m = \frac{-A}{B} = \tan\theta = \frac{y_2 - y_1}{x_2 - x_1} \qquad 7.13$$

$$b = \frac{-C}{B} \qquad 7.14$$

$$a = \frac{-C}{A} \qquad 7.15$$

- *point-slope form*

$$y - y_1 = m(x - x_1) \qquad 7.16$$

- *intercept form*

$$\frac{x}{a} + \frac{y}{b} = 1 \qquad 7.17$$

- *two-point form*

$$\frac{y - y_1}{x - x_1} = \frac{y_2 - y_1}{x_2 - x_1} \qquad 7.18$$

- *normal form*

$$x\cos\beta + y\sin\beta - d = 0 \qquad 7.19$$

(d and β are constants; x and y are variables.)

- *polar form*

$$r = \frac{d}{\cos(\beta - \alpha)} \qquad 7.20$$

(d and β are constants; r and α are variables.)

Figure 7.8 *Straight Line*

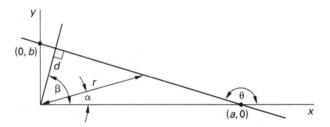

14. DISTANCES BETWEEN GEOMETRIC FIGURES

- Between two points in (x, y, z) format,

$$d = \sqrt{(x_2 - x_1)^2 + (y_2 - y_1)^2 + (z_2 - z_1)^2} \qquad 7.48$$

- Between a point (x_0, y_0) and a line $Ax + By + C = 0$,

$$d = \frac{|Ax_0 + By_0 + C|}{\sqrt{A^2 + B^2}} \qquad 7.49$$

- Between a point (x_0, y_0, z_0) and a plane $Ax + By + Cz + D = 0$,

$$d = \frac{|Ax_0 + By_0 + Cz_0 + D|}{\sqrt{A^2 + B^2 + C^2}} \qquad 7.50$$

- Between two parallel lines $Ax + By + C = 0$,

$$d = \left|\frac{|C_2|}{\sqrt{A_2^2 + B_2^2}} - \frac{|C_1|}{\sqrt{A_1^2 + B_1^2}}\right| \qquad 7.51$$

17. CIRCLE

$$Ax^2 + Ay^2 + Dx + Ey + F = 0 \qquad 7.66$$

$$(x - h)^2 + (y - k)^2 = r^2 \qquad 7.67$$

18. PARABOLA

$$(y - k)^2 = 4p(x - h)\Big|_{\text{opens horizontally}} \qquad 7.73$$

$$y^2 = 4px\Big|_{\substack{\text{vertex at origin} \\ h = k = 0}} \qquad 7.74$$

$$(x - h)^2 = 4p(y - k)\Big|_{\text{opens vertically}} \qquad 7.75$$

$$x^2 = 4py\Big|_{\text{vertex at origin}} \qquad 7.76$$

Figure 7.13 Parabola

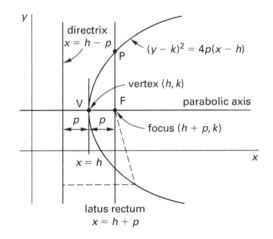

CERM Chapter 10
Differential Equations

16. APPLICATION: MIXING

$$m'(t) = \text{rate of addition} - \text{rate of removal}$$
$$= a(t) - r(t) \qquad 10.46$$

$$r(t) = c(t)o(t) \qquad 10.47$$

$$c(t) = \frac{m(t)}{V(t)} \qquad 10.48$$

$$m'(t) = a(t) - \frac{m(t)o(t)}{V(t)} \qquad 10.49$$

Figure 10.1 Fluid Mixture Problem

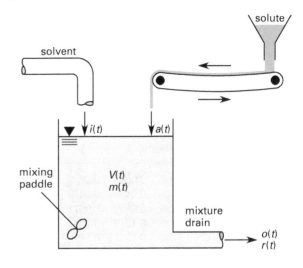

CERM Chapter 11
Probability and Statistical Analysis of Data

9. BINOMIAL DISTRIBUTION

$$p\{x\} = f(x) = \binom{n}{x}\hat{p}^x\hat{q}^{n-x} \qquad 11.28$$

$$\binom{n}{x} = \frac{n!}{(n - x)!x!} \qquad 11.29$$

12. POISSON DISTRIBUTION

$$p\{x\} = f(x) = \frac{e^{-\lambda}\lambda^x}{x!} \qquad [\lambda > 0] \qquad 11.34$$

λ is both the mean and the variance of the Poisson distribution.

15. NORMAL DISTRIBUTION

(See the "Areas Under the Standard Normal Curve" table.)

$$z = \frac{x_0 - \mu}{\sigma} \qquad 11.45$$

16. STUDENT'S *t*-DISTRIBUTION

(See the "Values of t_C for Student's *t*-Distribution" table.)

$$\text{df} = n - 1 \qquad 11.46$$

$$t = \frac{\overline{x} - \mu}{\dfrac{s}{\sqrt{n}}} \qquad 11.47$$

$$\text{UCL, LCL} = \overline{x} \pm t_{\alpha/2, n-1}\frac{s}{\sqrt{n}} \qquad 11.48$$

Areas Under the Standard Normal Curve

(0 to z)

z	0	1	2	3	4	5	6	7	8	9
0.0	0.0000	0.0040	0.0080	0.0120	0.0160	0.0199	0.0239	0.0279	0.0319	0.0359
0.1	0.0398	0.0438	0.0478	0.0517	0.0557	0.0596	0.0636	0.0675	0.0714	0.0754
0.2	0.0793	0.0832	0.0871	0.0910	0.0948	0.0987	0.1026	0.1064	0.1103	0.1141
0.3	0.1179	0.1217	0.1255	0.1293	0.1331	0.1368	0.1406	0.1443	0.1480	0.1517
0.4	0.1554	0.1591	0.1628	0.1664	0.1700	0.1736	0.1772	0.1808	0.1844	0.1879
0.5	0.1915	0.1950	0.1985	0.2019	0.2054	0.2088	0.2123	0.2157	0.2190	0.2224
0.6	0.2258	0.2291	0.2324	0.2357	0.2389	0.2422	0.2454	0.2486	0.2518	0.2549
0.7	0.2580	0.2612	0.2642	0.2673	0.2704	0.2734	0.2764	0.2794	0.2823	0.2852
0.8	0.2881	0.2910	0.2939	0.2967	0.2996	0.3023	0.3051	0.3078	0.3106	0.3133
0.9	0.3159	0.3186	0.3212	0.3238	0.3264	0.3289	0.3315	0.3340	0.3365	0.3389
1.0	0.3413	0.3438	0.3461	0.3485	0.3508	0.3531	0.3554	0.3577	0.3599	0.3621
1.1	0.3643	0.3665	0.3686	0.3708	0.3729	0.3749	0.3770	0.3790	0.3810	0.3830
1.2	0.3849	0.3869	0.3888	0.3907	0.3925	0.3944	0.3962	0.3980	0.3997	0.4015
1.3	0.4032	0.4049	0.4066	0.4082	0.4099	0.4115	0.4131	0.4147	0.4162	0.4177
1.4	0.4192	0.4207	0.4222	0.4236	0.4251	0.4265	0.4279	0.4292	0.4306	0.4319
1.5	0.4332	0.4345	0.4357	0.4370	0.4382	0.4394	0.4406	0.4418	0.4429	0.4441
1.6	0.4452	0.4463	0.4474	0.4484	0.4495	0.4505	0.4515	0.4525	0.4535	0.4545
1.7	0.4554	0.4564	0.4573	0.4582	0.4591	0.4599	0.4608	0.4616	0.4625	0.4633
1.8	0.4641	0.4649	0.4656	0.4664	0.4671	0.4678	0.4686	0.4693	0.4699	0.4706
1.9	0.4713	0.4719	0.4726	0.4732	0.4738	0.4744	0.4750	0.4756	0.4761	0.4767
2.0	0.4772	0.4778	0.4783	0.4788	0.4793	0.4798	0.4803	0.4808	0.4812	0.4817
2.1	0.4821	0.4826	0.4830	0.4834	0.4838	0.4842	0.4846	0.4850	0.4854	0.4857
2.2	0.4861	0.4864	0.4868	0.4871	0.4875	0.4878	0.4881	0.4884	0.4887	0.4890
2.3	0.4893	0.4896	0.4898	0.4901	0.4904	0.4906	0.4909	0.4911	0.4913	0.4916
2.4	0.4918	0.4920	0.4922	0.4925	0.4927	0.4929	0.4931	0.4932	0.4934	0.4936
2.5	0.4938	0.4940	0.4941	0.4943	0.4945	0.4946	0.4948	0.4949	0.4951	0.4952
2.6	0.4953	0.4955	0.4956	0.4957	0.4959	0.4960	0.4961	0.4962	0.4963	0.4964
2.7	0.4965	0.4966	0.4967	0.4968	0.4969	0.4970	0.4971	0.4972	0.4973	0.4974
2.8	0.4974	0.4975	0.4976	0.4977	0.4977	0.4978	0.4979	0.4979	0.4980	0.4981
2.9	0.4981	0.4982	0.4982	0.4983	0.4984	0.4984	0.4985	0.4985	0.4986	0.4986
3.0	0.4987	0.4987	0.4987	0.4988	0.4988	0.4989	0.4989	0.4989	0.4990	0.4990
3.1	0.4990	0.4991	0.4991	0.4991	0.4992	0.4992	0.4992	0.4992	0.4993	0.4993
3.2	0.4993	0.4993	0.4994	0.4994	0.4994	0.4994	0.4994	0.4995	0.4995	0.4995
3.3	0.4995	0.4995	0.4996	0.4996	0.4996	0.4996	0.4996	0.4996	0.4996	0.4997
3.4	0.4997	0.4997	0.4997	0.4997	0.4997	0.4997	0.4997	0.4997	0.4997	0.4998
3.5	0.4998	0.4998	0.4998	0.4998	0.4998	0.4998	0.4998	0.4998	0.4998	0.4998
3.6	0.4998	0.4998	0.4999	0.4999	0.4999	0.4999	0.4999	0.4999	0.4999	0.4999
3.7	0.4999	0.4999	0.4999	0.4999	0.4999	0.4999	0.4999	0.4999	0.4999	0.4999
3.8	0.4999	0.4999	0.4999	0.4999	0.4999	0.4999	0.4999	0.4999	0.4999	0.4999
3.9	0.5000	0.5000	0.5000	0.5000	0.5000	0.5000	0.5000	0.5000	0.5000	0.5000

Values of t_C for Student's t-Distribution

(ν degrees of freedom; confidence level C; $\alpha = 1 - C$; shaded area $= p$)

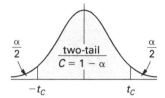

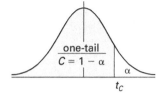

ν (df)	two-tail, t_C									
	$t_{99\%}$	$t_{98\%}$	$t_{95\%}$	$t_{90\%}$	$t_{80\%}$	$t_{60\%}$	$t_{50\%}$	$t_{40\%}$	$t_{20\%}$	$t_{10\%}$
	one-tail, t_C									
	$t_{99.5\%}$	$t_{99\%}$	$t_{97.5\%}$	$t_{95\%}$	$t_{90\%}$	$t_{80\%}$	$t_{75\%}$	$t_{70\%}$	$t_{60\%}$	$t_{55\%}$
1	63.657	31.821	12.706	6.314	3.078	1.376	1.000	0.727	0.325	0.158
2	9.925	6.965	4.303	2.920	1.886	1.061	0.816	0.617	0.289	0.142
3	5.841	4.541	3.182	2.353	1.638	0.978	0.765	0.584	0.277	0.137
4	4.604	3.747	2.776	2.132	1.533	0.941	0.741	0.569	0.271	0.134
5	4.032	3.365	2.571	2.015	1.476	0.920	0.727	0.559	0.267	0.132
6	3.707	3.143	2.447	1.943	1.440	0.906	0.718	0.553	0.265	0.131
7	3.499	2.998	2.365	1.895	1.415	0.896	0.711	0.549	0.263	0.130
8	3.355	2.896	2.300	1.860	1.397	0.889	0.706	0.546	0.262	0.130
9	3.250	2.821	2.262	1.833	1.383	0.883	0.703	0.543	0.261	0.129
10	3.169	2.764	2.228	1.812	1.372	0.879	0.700	0.542	0.260	0.129
11	3.106	2.718	2.201	1.796	1.363	0.876	0.697	0.540	0.260	0.129
12	3.055	2.681	2.179	1.782	1.356	0.873	0.695	0.539	0.259	0.128
13	3.012	2.650	2.160	1.771	1.350	0.870	0.694	0.538	0.259	0.128
14	2.977	2.624	2.145	1.761	1.345	0.868	0.692	0.537	0.258	0.128
15	2.947	2.602	2.131	1.753	1.341	0.866	0.691	0.536	0.258	0.128
16	2.921	2.583	2.120	1.746	1.337	0.865	0.690	0.535	0.258	0.128
17	2.898	2.567	2.110	1.740	1.333	0.863	0.689	0.534	0.257	0.128
18	2.878	2.552	2.101	1.734	1.330	0.862	0.688	0.534	0.257	0.127
19	2.861	2.539	2.093	1.729	1.328	0.861	0.688	0.533	0.257	0.127
20	2.845	2.528	2.086	1.725	1.325	0.860	0.687	0.533	0.257	0.127
21	2.831	2.518	2.080	1.721	1.323	0.859	0.686	0.532	0.257	0.127
22	2.819	2.508	2.074	1.717	1.321	0.858	0.686	0.532	0.256	0.127
23	2.807	2.500	2.069	1.714	1.319	0.858	0.685	0.532	0.256	0.127
24	2.797	2.492	2.064	1.711	1.318	0.857	0.685	0.531	0.256	0.127
25	2.787	2.485	2.060	1.708	1.316	0.856	0.684	0.531	0.256	0.127
26	2.779	2.479	2.056	1.706	1.315	0.856	0.684	0.531	0.256	0.127
27	2.771	2.473	2.052	1.703	1.314	0.855	0.684	0.531	0.256	0.127
28	2.763	2.467	2.048	1.701	1.313	0.855	0.683	0.530	0.256	0.127
29	2.756	2.462	2.045	1.699	1.311	0.854	0.683	0.530	0.256	0.127
30	2.750	2.457	2.042	1.697	1.310	0.854	0.683	0.530	0.256	0.127
40	2.705	2.423	2.021	1.684	1.303	0.851	0.681	0.529	0.255	0.126
60	2.660	2.390	2.000	1.671	1.296	0.848	0.679	0.527	0.254	0.126
120	2.617	2.358	1.980	1.658	1.289	0.845	0.677	0.526	0.254	0.126
∞	2.557	2.326	1.960	1.645	1.282	0.842	0.674	0.524	0.253	0.126

20. APPLICATION: RELIABILITY

$$R\{t\} = e^{-\lambda t} = e^{-t/\text{MTTF}} \qquad 11.59$$

$$\lambda = \frac{1}{\text{MTTF}} \qquad 11.60$$

$$R\{t\} = 1 - F(t) = 1 - \left(1 - e^{-\lambda t}\right) = e^{-\lambda t} \qquad 11.61$$

$$z\{t\} = \lambda \qquad 11.62$$

23. MEASURES OF CENTRAL TENDENCY

$$\overline{x} = \left(\frac{1}{n}\right)(x_1 + x_2 + \cdots + x_n) = \frac{\sum x_i}{n} \qquad 11.68$$

$$\text{geometric mean} = \sqrt[n]{x_1 x_2 x_3 \cdots x_n} \quad [x_i > 0] \qquad 11.69$$

$$\text{harmonic mean} = \frac{n}{\dfrac{1}{x_1} + \dfrac{1}{x_2} + \cdots + \dfrac{1}{x_n}} \qquad 11.70$$

$$x_{\text{rms}} = \sqrt{\frac{\sum x_i^2}{n}} \qquad 11.71$$

24. MEASURES OF DISPERSION

$$\sigma = \sqrt{\frac{\sum (x_i - \mu)^2}{N}} = \sqrt{\frac{\sum x_i^2}{N} - \mu^2} \qquad 11.74$$

$$s = \sqrt{\frac{\sum (x_i - \overline{x})^2}{n-1}} = \sqrt{\frac{\sum x_i^2 - \dfrac{\left(\sum x_i\right)^2}{n}}{n-1}} \qquad 11.75$$

$$\sigma_{\text{sample}} = s\sqrt{\frac{n-1}{n}} \qquad 11.76$$

$$\text{coefficient of variation} = \frac{s}{\overline{x}} \qquad 11.77$$

30. APPLICATION: CONFIDENCE LIMITS

$$\text{LCL}: \mu - z\sigma$$

$$\text{UCL}: \mu + z\sigma$$

$$p\{\overline{x} > L\} = p\left\{ z > \left| \frac{L - \mu}{\dfrac{\sigma}{\sqrt{n}}} \right| \right\} \qquad 11.86$$

$$p\{\text{LCL} < \overline{x} < \text{UCL}\} = p\left\{ \left| \frac{\text{LCL} - \mu}{\dfrac{\sigma}{\sqrt{n}}} \right| < z < \left| \frac{\text{UCL} - \mu}{\dfrac{\sigma}{\sqrt{n}}} \right| \right\} \qquad 11.87$$

Table 11.2 *Values of z for Various Confidence Levels*

confidence level, C	one-tail limit, z	two-tail limit, z
90%	1.28	1.645
95%	1.645	1.96
97.5%	1.96	2.24
99%	2.33	2.575
99.5%	2.575	2.81
99.75%	2.81	3.00

CERM Chapter 13
Energy, Work, and Power

3. WORK

$$W_{\text{constant force}} = \mathbf{F} \cdot \mathbf{s} = Fs \cos \phi \quad [\text{linear systems}] \qquad 13.6$$

$$\begin{aligned} W_{\text{constant torque}} &= \mathbf{T} \cdot \theta \\ &= Fr\theta \cos \phi \quad [\text{rotational systems}] \qquad 13.7 \end{aligned}$$

$$W_{\text{friction}} = F_f s \qquad 13.8$$

$$W_{\text{gravity}} = mg(h_2 - h_1) \qquad [\text{SI}] \quad 13.9(a)$$

$$W_{\text{gravity}} = \frac{mg(h_2 - h_1)}{g_c} \qquad [\text{U.S.}] \quad 13.9(b)$$

$$W_{\text{spring}} = \tfrac{1}{2}k(\delta_2^2 - \delta_1^2) \qquad 13.10$$

Figure 13.1 *Work of a Constant Force*

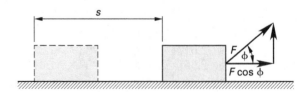

4. POTENTIAL ENERGY OF A MASS

$$E_{\text{potential}} = mgh \qquad \text{[SI]} \qquad 13.11(a)$$

$$E_{\text{potential}} = \frac{mgh}{g_c} \qquad \text{[U.S.]} \qquad 13.11(b)$$

5. KINETIC ENERGY OF A MASS

$$E_{\text{kinetic}} = \tfrac{1}{2}m\text{v}^2 \qquad \text{[SI]} \qquad 13.13(a)$$

$$E_{\text{kinetic}} = \frac{m\text{v}^2}{2g_c} \qquad \text{[U.S.]} \qquad 13.13(b)$$

$$E_{\text{rotational}} = \tfrac{1}{2}I\omega^2 \qquad \text{[SI]} \qquad 13.14(a)$$

$$E_{\text{rotational}} = \frac{I\omega^2}{2g_c} \qquad \text{[U.S.]} \qquad 13.14(b)$$

6. SPRING ENERGY

$$E_{\text{spring}} = \tfrac{1}{2}k\delta^2 \qquad 13.16$$

7. PRESSURE ENERGY OF A MASS

$$E_{\text{flow}} = \frac{mp}{\rho} = mpv \qquad [v = \text{specific volume}] \qquad 13.17$$

8. INTERNAL ENERGY OF A MASS

$$U_2 - U_1 = Q \qquad 13.18$$

$$Q = mc_v\Delta T \qquad [\text{constant-volume process}] \qquad 13.21$$

$$Q = mc_p\Delta T \qquad [\text{constant-pressure process}] \qquad 13.22$$

9. WORK-ENERGY PRINCIPLE

$$W = \Delta E = E_2 - E_1 \qquad 13.23$$

11. POWER

$$P = \frac{W}{\Delta t} \qquad 13.24$$

$$P = F\text{v} \qquad [\text{linear systems}] \qquad 13.25$$

$$P = T\omega \qquad [\text{rotational systems}] \qquad 13.26$$

For flowing fluid,

$$P = \dot{m}\Delta u \qquad 13.27$$

Table 13.2 Useful Power Conversion Formulas

$$
\begin{aligned}
1 \text{ hp} &= 550 \text{ ft-lbf/sec} \\
&= 33{,}000 \text{ ft-lbf/min} \\
&= 0.7457 \text{ kW} \\
&= 0.7068 \text{ Btu/sec} \\
1 \text{ kW} &= 737.6 \text{ ft-lbf/sec} \\
&= 44{,}250 \text{ ft-lbf/min} \\
&= 1.341 \text{ hp} \\
&= 0.9483 \text{ Btu/sec} \\
1 \text{ Btu/sec} &= 778.17 \text{ ft-lbf/sec} \\
&= 46{,}680 \text{ ft-lbf/min} \\
&= 1.415 \text{ hp}
\end{aligned}
$$

12. EFFICIENCY

$$\eta = \frac{P_{\text{ideal}}}{P_{\text{actual}}} \qquad [P_{\text{actual}} \geq P_{\text{ideal}}] \qquad 13.28$$

$$\eta = \frac{P_{\text{actual}}}{P_{\text{ideal}}} \qquad [P_{\text{ideal}} \geq P_{\text{actual}}] \qquad 13.29$$

Water Resources

Fluid Properties

4. DENSITY

- *gases*

$$\rho = \frac{p}{RT} \qquad \qquad 14.4$$

5. SPECIFIC VOLUME

$$v = \frac{1}{\rho} \qquad \qquad 14.5$$

6. SPECIFIC GRAVITY

$$\text{SG}_{\text{liquid}} = \frac{\rho_{\text{liquid}}}{\rho_{\text{water}}} \qquad \qquad 14.6$$

$$\text{SG}_{\text{gas}} = \frac{\rho_{\text{gas}}}{\rho_{\text{air}}} \qquad \qquad 14.7$$

$$\text{SG}_{\text{gas}} = \frac{\text{MW}_{\text{gas}}}{\text{MW}_{\text{air}}} = \frac{\text{MW}_{\text{gas}}}{29.0} = \frac{R_{\text{air}}}{R_{\text{gas}}}$$

$$= \frac{53.3 \, \frac{\text{ft-lbf}}{\text{lbm-}^\circ \text{R}}}{R_{\text{gas}}} \qquad \qquad 14.8$$

7. SPECIFIC WEIGHT

$$\gamma = g\rho \qquad \text{[SI]} \qquad 14.13(a)$$

$$\gamma = \rho \times \frac{g}{g_c} \qquad \text{[U.S.]} \qquad 14.13(b)$$

9. VISCOSITY

The constant of proportionality, μ, is the absolute viscosity.

$$\frac{F}{A} = \mu \frac{d\text{v}}{dy} \qquad \qquad 14.18$$

$$\tau = \mu \frac{d\text{v}}{dy} \qquad \qquad 14.19$$

10. KINEMATIC VISCOSITY

$$\nu = \frac{\mu}{\rho} \qquad \text{[SI]} \qquad 14.20(a)$$

$$\nu = \frac{\mu g_c}{\rho} = \frac{\mu g}{\gamma} \qquad \text{[U.S.]} \qquad 14.20(b)$$

19. BULK MODULUS

The bulk modulus, E, is the reciprocal of compressibility, β.

$$E = \frac{1}{\beta} \qquad \qquad 14.37$$

20. SPEED OF SOUND

- *liquids*

$$a = \sqrt{\frac{E}{\rho}} = \sqrt{\frac{1}{\beta\rho}} \qquad \text{[SI]} \qquad 14.38(a)$$

$$a = \sqrt{\frac{Eg_c}{\rho}} = \sqrt{\frac{g_c}{\beta\rho}} \qquad \text{[U.S.]} \qquad 14.38(b)$$

- *gases*

$$a = \sqrt{\frac{E}{\rho}} = \sqrt{\frac{kp}{\rho}}$$

$$= \sqrt{kRT} = \sqrt{\frac{kR^*T}{\text{MW}}} \qquad \text{[SI]} \qquad 14.39(a)$$

$$a = \sqrt{\frac{Eg_c}{\rho}} = \sqrt{\frac{kg_c p}{\rho}}$$

$$= \sqrt{kg_c RT} = \sqrt{\frac{kg_c R^*T}{\text{MW}}} \qquad \text{[U.S.]} \qquad 14.39(b)$$

CERM Chapter 15
Fluid Statics

2. MANOMETERS

$$p_2 - p_1 = \rho_m g h \qquad \text{[SI]} \quad 15.3(a)$$

$$p_2 - p_1 = \rho_m h \times \frac{g}{g_c} = \gamma_m h \qquad \text{[U.S.]} \quad 15.3(b)$$

$$p_2 - p_1 = g(\rho_m h + \rho_1 h_1 - \rho_2 h_2) \qquad \text{[SI]} \quad 15.4(a)$$

$$p_2 - p_1 = (\rho_m h + \rho_1 h_1 - \rho_2 h_2) \times \frac{g}{g_c}$$

$$= \gamma_m h + \gamma_1 h_1 - \gamma_2 h_2 \qquad \text{[U.S.]} \quad 15.4(b)$$

Figure 15.4 Manometer Requiring Corrections

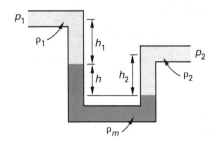

3. HYDROSTATIC PRESSURE

$$p = \frac{F}{A} \qquad\qquad 15.5$$

4. FLUID HEIGHT EQUIVALENT TO PRESSURE

$$p = \rho g h \qquad \text{[SI]} \quad 15.6(a)$$

$$p = \frac{\rho g h}{g_c} = \gamma h \qquad \text{[U.S.]} \quad 15.6(b)$$

7. PRESSURE ON A RECTANGULAR VERTICAL PLANE SURFACE

$$\overline{p} = \tfrac{1}{2}(p_1 + p_2) \qquad\qquad 15.13$$

$$\overline{p} = \tfrac{1}{2}\rho g(h_1 + h_2) \qquad \text{[SI]} \quad 15.14(a)$$

$$\overline{p} = \frac{\tfrac{1}{2}\rho g(h_1 + h_2)}{g_c} = \tfrac{1}{2}\gamma(h_1 + h_2) \qquad \text{[U.S.]} \quad 15.14(b)$$

$$R = \overline{p}A \qquad\qquad 15.15$$

$$h_R = \frac{2}{3}\left(h_1 + h_2 - \frac{h_1 h_2}{h_1 + h_2}\right) \qquad 15.16$$

Figure 15.9 Hydrostatic Pressure on a Vertical Plane Surface

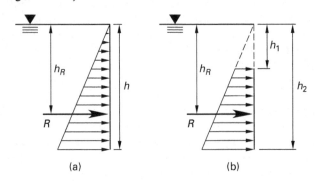

(a)   (b)

9. PRESSURE ON A GENERAL PLANE SURFACE

$$\overline{p} = \rho g h_c \sin\theta \qquad \text{[SI]} \quad 15.21(a)$$

$$\overline{p} = \frac{\rho g h_c \sin\theta}{g_c} = \gamma h_c \sin\theta \qquad \text{[U.S.]} \quad 15.21(b)$$

$$R = \overline{p}A \qquad\qquad 15.22$$

$$h_R = h_c + \frac{I_c}{A h_c} \qquad\qquad 15.23$$

Figure 15.11 General Plane Surface

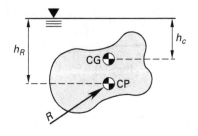

13. HYDROSTATIC FORCES ON A DAM

$$M_{\text{overturning}} = R_x y_{R_x} \qquad\qquad 15.28$$

$$M_{\text{resisting}} = R_y x_{R_y} + W x_{\text{CG}} \qquad\qquad 15.29$$

$$(\text{FS})_{\text{overturning}} = \frac{M_{\text{resisting}}}{M_{\text{overturning}}} \qquad\qquad 15.30$$

$$F_f = \mu_{\text{static}} N = \mu_{\text{static}}(W + R_y) \qquad\qquad 15.31$$

$$(\text{FS})_{\text{sliding}} = \frac{F_f}{R_x} \qquad\qquad 15.32$$

$$p_{\max}, p_{\min} = \left(\frac{R_y + W}{b}\right)\left(1 \pm \frac{6e}{b}\right) \quad \text{[per unit width]}$$

$$15.33$$

$$e = \frac{b}{2} - x_v \qquad 15.34$$

$$x_v = \frac{M_{\text{resisting}} - M_{\text{overturning}}}{R_y + W} \qquad 15.35$$

Figure 15.16 Dam

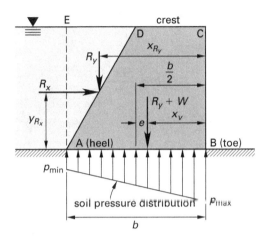

18. BUOYANCY

$$F_{\text{buoyant}} = \rho g V_{\text{displaced}} \qquad \text{[SI]} \quad 15.53(a)$$

$$F_{\text{buoyant}} = \frac{\rho g V_{\text{displaced}}}{g_c} = \gamma V_{\text{displaced}} \qquad \text{[U.S.]} \quad 15.53(b)$$

$$\text{SG} = \frac{W_{\text{dry}}}{W_{\text{dry}} - W_{\text{submerged}}} \qquad 15.54$$

CERM Chapter 16
Fluid Flow Parameters

2. KINETIC ENERGY

$$E_{\text{v}} = \frac{\text{v}^2}{2} \qquad \text{[SI]} \quad 16.1(a)$$

$$E_{\text{v}} = \frac{\text{v}^2}{2g_c} \qquad \text{[U.S.]} \quad 16.1(b)$$

3. POTENTIAL ENERGY

$$E_z = zg \qquad \text{[SI]} \quad 16.2(a)$$

$$E_z = \frac{zg}{g_c} \qquad \text{[U.S.]} \quad 16.2(b)$$

4. PRESSURE ENERGY

$$E_p = \frac{p}{\rho} \qquad 16.3$$

5. BERNOULLI EQUATION

$$E_t = \frac{p}{\rho} + \frac{\text{v}^2}{2} + zg \qquad \text{[SI]} \quad 16.5(a)$$

$$E_t = \frac{p}{\rho} + \frac{\text{v}^2}{2g_c} + \frac{zg}{g_c} \qquad \text{[U.S.]} \quad 16.5(b)$$

8. HYDRAULIC RADIUS

$$r_h = \frac{\text{area in flow}}{\text{wetted perimeter}} = \frac{A}{s} \qquad 16.12$$

9. HYDRAULIC DIAMETER

$$D_h = 4r_h \qquad 16.14$$

Table 16.1 Hydraulic Diameters for Common Conduit Shapes

conduit cross section	D_h
flowing full	
circle	D
annulus (outer diameter D_o, inner diameter D_i)	$D_o - D_i$
square (side L)	L
rectangle (sides L_1 and L_2)	$\dfrac{2L_1 L_2}{L_1 + L_2}$
flowing partially full	
half-filled circle (diameter D)	D
rectangle (h deep, L wide)	$\dfrac{4hL}{L + 2h}$
wide, shallow stream (h deep)	$4h$
triangle, vertex down (h deep, L broad, s side)	$\dfrac{hL}{s}$
trapezoid (h deep, a wide at top, b wide at bottom, s side)	$\dfrac{2h(a + b)}{b + 2s}$

10. REYNOLDS NUMBER

$$\text{Re} = \frac{D_h \text{v} \rho}{\mu} \qquad \text{[SI]} \quad 16.16(a)$$

$$\text{Re} = \frac{D_h \text{v} \rho}{g_c \mu} \qquad \text{[U.S.]} \quad 16.16(b)$$

$$\text{Re} = \frac{D_h \text{v}}{\nu} \qquad 16.17$$

15. ENERGY GRADE LINE

$$\text{elevation of EGL} = h_p + h_v + h_z \qquad 16.28$$
$$\text{elevation of HGL} = h_p + h_z \qquad 16.29$$

16. SPECIFIC ENERGY

$$E_{\text{specific}} = E_p + E_v \qquad 16.31$$

$$E_{\text{specific}} = \frac{p}{\rho} + \frac{v^2}{2} \qquad \text{[SI]} \quad 16.32(a)$$

$$E_{\text{specific}} = \frac{p}{\rho} + \frac{v^2}{2g_c} \qquad \text{[U.S.]} \quad 16.32(b)$$

$$E_{\text{specific}} = d + \frac{v^2}{2g} \qquad \text{[open channel]} \qquad 16.33$$

CERM Chapter 17
Fluid Dynamics

2. CONSERVATION OF MASS

$$\rho_1 A_1 v_1 = \rho_2 A_2 v_2 \qquad 17.2$$

7. FRICTION FACTOR

(See Fig. 17.4.)

9. ENERGY LOSS DUE TO FRICTION: TURBULENT FLOW

- *Darcy equation*

$$h_f = \frac{fLv^2}{2Dg} \qquad 17.27$$

- *Hazen-Williams equation*

$$h_{f,\text{ft}} = \frac{3.022 v_{\text{ft/sec}}^{1.85} L_{\text{ft}}}{C^{1.85} D_{\text{ft}}^{1.17}} = \frac{10.44 L_{\text{ft}} Q_{\text{gpm}}^{1.85}}{C^{1.85} d_{\text{in}}^{4.87}} \qquad \text{[U.S.]} \quad 17.29$$

Figure 17.4 *Moody Friction Factor Chart*

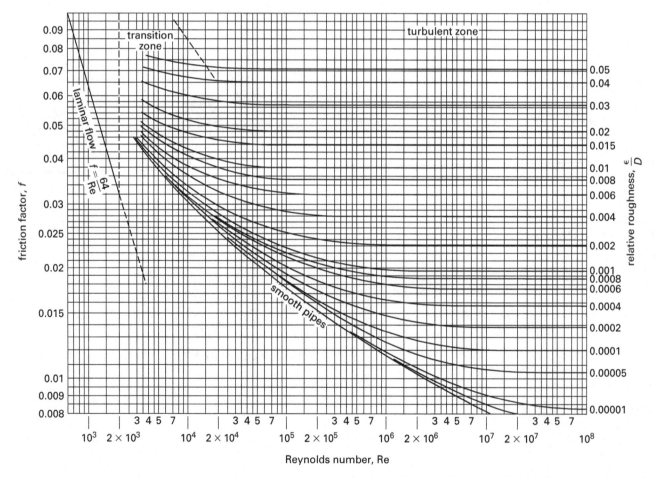

Reprinted with permission from L. F. Moody, "Friction Factor for Pipe Flow," *ASME Transactions*, Vol. 66, published by the American Society of Mechanical Engineers, copyright © 1944.

15. MINOR LOSSES

$$h_m = K h_{\mathrm{v}} \qquad 17.39$$

$$K = \frac{f L_e}{D} \qquad 17.40$$

- *sudden enlargements* (D_1 is the smaller of the two diameters)

$$K = \left(1 - \left(\frac{D_1}{D_2}\right)^2\right)^2 \qquad 17.41$$

- *sudden contractions* (D_1 is the smaller of the two diameters)

$$K = \tfrac{1}{2}\left(1 - \left(\frac{D_1}{D_2}\right)^2\right) \qquad 17.42$$

16. VALVE FLOW COEFFICIENTS

$$\dot{V}_{\mathrm{m^3/h}} = K_v \sqrt{\frac{\Delta p_{\mathrm{bars}}}{\mathrm{SG}}} \qquad \text{[SI]} \quad 17.48(a)$$

$$Q_{\mathrm{gpm}} = C_v \sqrt{\frac{\Delta p_{\mathrm{psi}}}{\mathrm{SG}}} \qquad \text{[U.S.]} \quad 17.48(b)$$

21. DISCHARGE FROM TANKS

$$\mathrm{v}_t = \sqrt{2gh} \qquad 17.66$$

$$h = z_1 - z_2 \qquad 17.67$$

$$\mathrm{v}_o = C_{\mathrm{v}}\sqrt{2gh} \qquad 17.68$$

$$C_{\mathrm{v}} = \frac{\text{actual velocity}}{\text{theoretical velocity}} = \frac{\mathrm{v}_o}{\mathrm{v}_t} \qquad 17.69$$

$$C_c = \frac{A_{\mathrm{vena\,contracta}}}{A_o} \qquad 17.73$$

$$C_d = C_{\mathrm{v}} C_c = \frac{\dot{V}}{\dot{V}_t} \qquad 17.75$$

Figure 17.10 Discharge from a Tank

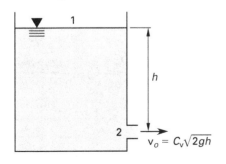

23. COORDINATES OF A FLUID STREAM

$$\mathrm{v}_x = \mathrm{v}_o \qquad \text{[horizontal discharge]} \qquad 17.77$$

$$x = \mathrm{v}_o t = \mathrm{v}_o \sqrt{\frac{2y}{g}} = 2 C_{\mathrm{v}}\sqrt{hy} \qquad 17.78$$

$$\mathrm{v}_y = gt \qquad 17.79$$

$$y = \frac{gt^2}{2} = \frac{gx^2}{2\mathrm{v}_o^2} = \frac{x^2}{4hC_{\mathrm{v}}^2} \qquad 17.80$$

Figure 17.13 Coordinates of a Fluid Stream

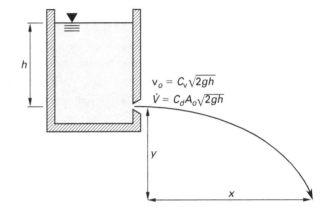

26. PRESSURE CULVERTS

$$\dot{V} = C_d A \mathrm{v} = C_d A \sqrt{2gh_{\mathrm{effective}}} \qquad 17.85$$

$$h_{\mathrm{effective}} = h - h_{f,\mathrm{barrel}} - h_{m,\mathrm{entrance}} \qquad 17.86$$

Figure 17.14 Simple Pipe Culvert

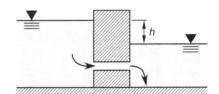

33. PITOT-STATIC GAUGE

$$\frac{\mathrm{v}^2}{2} = \frac{p_t - p_s}{\rho} = \frac{h(\rho_m - \rho)g}{\rho} \qquad \text{[SI]} \quad 17.144(a)$$

$$\frac{\mathrm{v}^2}{2g_c} = \frac{p_t - p_s}{\rho} = \frac{h(\rho_m - \rho)}{\rho} \times \frac{g}{g_c} \qquad \text{[U.S.]} \quad 17.144(b)$$

$$\mathrm{v} = \sqrt{\frac{2gh(\rho_m - \rho)}{\rho}} \qquad 17.145$$

Figure 17.23 *Pitot-Static Gauge*

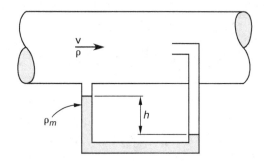

Figure 17.25 *Venturi Meter with Manometer*

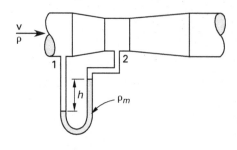

34. VENTURI METER

$$v_2 = C_v v_{2,\text{ideal}}$$

$$= \left(\frac{C_v}{\sqrt{1 - \left(\frac{A_2}{A_1} \right)^2}} \right) \sqrt{\frac{2(p_1 - p_2)}{\rho}} \quad \text{[SI]} \quad 17.150(a)$$

$$v_2 = C_v v_{2,\text{ideal}}$$

$$= \left(\frac{C_v}{\sqrt{1 - \left(\frac{A_2}{A_1} \right)^2}} \right) \sqrt{\frac{2g_c(p_1 - p_2)}{\rho}} \quad \text{[U.S.]} \quad 17.150(b)$$

$$\beta = \frac{D_2}{D_1} \qquad 17.151$$

$$F_{\text{va}} = \frac{1}{\sqrt{1 - \left(\frac{A_2}{A_1} \right)^2}} = \frac{1}{\sqrt{1 - \beta^4}} \qquad 17.152$$

$$v_2 = C_v v_{2,\text{ideal}} = \left(\frac{C_v}{\sqrt{1 - \beta^4}} \right) \sqrt{\frac{2g(\rho_m - \rho)h}{\rho}}$$

$$= C_v F_{\text{va}} \sqrt{\frac{2g(\rho_m - \rho)h}{\rho}} \qquad 17.153$$

$$\dot{V} = C_d A_2 v_{2,\text{ideal}} \qquad 17.154$$

$$C_f = C_d F_{\text{va}} = \frac{C_d}{\sqrt{1 - \beta^4}} \qquad 17.155$$

$$\dot{V} = C_f A_2 \sqrt{\frac{2g(\rho_m - \rho)h}{\rho}} \qquad 17.156$$

35. ORIFICE METER

$$A_2 = C_c A_o \qquad 17.157$$

$$v_o = \left(\frac{C_d}{\sqrt{1 - \left(\frac{C_c A_o}{A_1} \right)^2}} \right) \sqrt{\frac{2(p_1 - p_2)}{\rho}} \quad \text{[SI]} \quad 17.158(a)$$

$$v_o = \left(\frac{C_d}{\sqrt{1 - \left(\frac{C_c A_o}{A_1} \right)^2}} \right) \sqrt{\frac{2g_c(p_1 - p_2)}{\rho}}$$

$$\text{[U.S.]} \quad 17.158(b)$$

$$v_o = \left(\frac{C_d}{\sqrt{1 - \left(\frac{C_c A_o}{A_1} \right)^2}} \right) \sqrt{\frac{2g(\rho_m - \rho)h}{\rho}} \qquad 17.159$$

$$F_{\text{va}} = \frac{1}{\sqrt{1 - \left(\frac{C_c A_o}{A_1} \right)^2}} \qquad 17.160$$

$$C_f = C_d F_{\text{va}} \qquad 17.161$$

$$\dot{V} = C_f A_o \sqrt{\frac{2g(\rho_m - \rho)h}{\rho}} = C_f A_o \sqrt{\frac{2(p_1 - p_2)}{\rho}}$$

$$\text{[SI]} \quad 17.162(a)$$

$$\dot{V} = C_f A_o \sqrt{\frac{2g(\rho_m - \rho)h}{\rho}} = C_f A_o \sqrt{\frac{2g_c(p_1 - p_2)}{\rho}}$$

$$\text{[U.S.]} \quad 17.162(b)$$

Figure 17.27 Orifice Meter with Differential Manometer

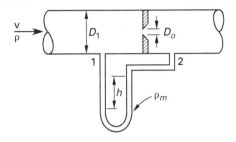

47. CONFINED STREAMS IN PIPE BENDS

$$F_x = p_2 A_2 \cos \theta - p_1 A_1 + \dot{m}(v_2 \cos \theta - v_1)$$
$$[\text{SI}] \quad 17.202(a)$$

$$F_x = p_2 A_2 \cos \theta - p_1 A_1 + \frac{\dot{m}(v_2 \cos \theta - v_1)}{g_c}$$
$$[\text{U.S.}] \quad 17.202(b)$$

$$F_y = (p_2 A_2 + \dot{m} v_2) \sin \theta \qquad [\text{SI}] \quad 17.203(a)$$

$$F_y = \left(p_2 A_2 + \frac{\dot{m} v_2}{g_c} \right) \sin \theta \qquad [\text{U.S.}] \quad 17.203(b)$$

Figure 17.40 Pipe Bend

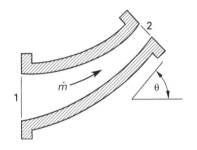

48. WATER HAMMER

- *one-way*

$$t = \frac{L}{a} \qquad 17.206$$

- *round trip*

$$t = \frac{2L}{a} \qquad 17.207$$

$$\Delta p = \rho a \Delta v \qquad [\text{SI}] \quad 17.208(a)$$

$$\Delta p = \frac{\rho a \Delta v}{g_c} \qquad [\text{U.S.}] \quad 17.208(b)$$

$$a = \sqrt{\frac{\dfrac{E_{\text{fluid}} t_{\text{pipe}} E_{\text{pipe}}}{t_{\text{pipe}} E_{\text{pipe}} + c_P D_{\text{pipe}} E_{\text{fluid}}}}{\rho}} \qquad 17.210$$

$$a = \sqrt{\frac{dp}{d\rho}} = \sqrt{\frac{E}{\rho}} \quad \begin{bmatrix} \text{inelastic pipe;} \\ \text{consistent units} \end{bmatrix} \qquad 17.205$$

Figure 17.41 Water Hammer

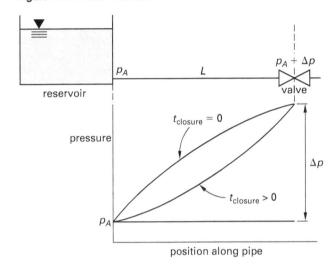

52. DRAG

$$F_D = \frac{C_D A \rho v^2}{2} \quad [\text{turbulent}] \qquad [\text{SI}] \quad 17.218(a)$$

$$F_D = \frac{C_D A \rho v^2}{2 g_c} \quad [\text{turbulent}] \qquad [\text{U.S.}] \quad 17.218(b)$$

53. DRAG ON SPHERES AND DISKS

$$C_D = \frac{24}{\text{Re}} \qquad [\text{Re} < 0.4] \qquad 17.219$$

$$F_D = 6\pi \mu v R = 3\pi \mu v D \quad [\text{laminar}] \qquad 17.220$$

54. TERMINAL VELOCITY

$$v_{\text{terminal}} = \sqrt{\frac{2(mg - F_b)}{C_D A \rho_{\text{fluid}}}} = \sqrt{\frac{2 V g (\rho_{\text{object}} - \rho_{\text{fluid}})}{C_D A \rho_{\text{fluid}}}} \qquad 17.222$$

- *sphere*

$$v_{\text{terminal,sphere}} = \sqrt{\frac{4 D g (\rho_{\text{sphere}} - \rho_{\text{fluid}})}{3 C_D \rho_{\text{fluid}}}} \qquad 17.223$$

- *small particles (Stokes' law)*

$$v_{terminal} = \frac{D^2 g(\rho_{sphere} - \rho_{fluid})}{18\mu} \quad \text{[laminar]}$$
$$\text{[SI]} \quad 17.224(a)$$

$$v_{terminal} = \frac{D^2(\rho_{sphere} - \rho_{fluid})}{18\mu} \times \frac{g}{g_c} \quad \text{[laminar]}$$
$$\text{[U.S.]} \quad 17.224(b)$$

58. SIMILARITY

$$L_r = \frac{\text{size of model}}{\text{size of prototype}} \qquad 17.229$$

59. VISCOUS AND INERTIAL FORCES DOMINATE

$$Re_m = Re_p \qquad 17.232$$

$$\frac{L_m v_m}{\nu_m} = \frac{L_p v_p}{\nu_p} \qquad 17.233$$

60. INERTIAL AND GRAVITATIONAL FORCES DOMINATE

$$Fr = \frac{v^2}{Lg} \qquad 17.234$$

$$Re_m = Re_p \qquad 17.235$$

$$Fr_m = Fr_p \qquad 17.236$$

$$\frac{\nu_m}{\nu_p} = \left(\frac{L_m}{L_p}\right)^{3/2} = L_r^{3/2} \qquad 17.237$$

$$n_r = L_r^{1/6} \qquad 17.238$$

61. SURFACE TENSION FORCE DOMINATES

$$We = \frac{v^2 L\rho}{\sigma} \qquad 17.239$$

$$We_m = We_p \qquad 17.240$$

CERM Chapter 18
Hydraulic Machines

6. CENTRIFUGAL PUMPS

$$v_{tip} = \frac{\pi D n}{60 \frac{sec}{min}} = \frac{D\omega}{2} \qquad 18.3$$

10. PUMPING POWER

Table 18.5 Hydraulic Horsepower Equations[a]

	Q (gal/min)	$\dot{m}$ (lbm/sec)	$\dot{V}$ (ft³/sec)
h_A in feet	$\frac{h_A Q(SG)}{3956}$	$\frac{h_A \dot{m}}{550} \times \frac{g}{g_c}$	$\frac{h_A \dot{V}(SG)}{8.814}$
Δp in psi[b]	$\frac{\Delta p Q}{1714}$	$\frac{\Delta p \dot{m}}{(238.3)(SG)} \times \frac{g}{g_c}$	$\frac{\Delta p \dot{V}}{3.819}$
Δp in psf[b]	$\frac{\Delta p Q}{2.468 \times 10^5}$	$\frac{\Delta p \dot{m}}{(34,320)(SG)} \times \frac{g}{g_c}$	$\frac{\Delta p \dot{V}}{550}$
W in $\frac{ft\text{-}lbf}{lbm}$	$\frac{W Q(SG)}{3956}$	$\frac{W \dot{m}}{550}$	$\frac{W \dot{V}(SG)}{8.814}$

(Multiply horsepower by 0.7457 to obtain kilowatts.)
[a]based on $\rho_{water} = 62.4$ lbm/ft³ and $g = 32.2$ ft/sec²
[b]Velocity head changes must be included in Δp.

Table 18.6 Hydraulic Kilowatt Equations[a]

	Q (L/s)	$\dot{m}$ (kg/s)	$\dot{V}$ (m³/s)
h_A in meters	$\frac{9.81 h_A Q(SG)}{1000}$	$\frac{9.81 h_A \dot{m}}{1000}$	$9.81 h_A \dot{V}(SG)$
Δp in kPa[b]	$\frac{\Delta p Q}{1000}$	$\frac{\Delta p \dot{m}}{1000(SG)}$	$\Delta p \dot{V}$
W in $\frac{J}{kg}$[b]	$\frac{W Q(SG)}{1000}$	$\frac{W \dot{m}}{1000}$	$W \dot{V}(SG)$

(Multiply kilowatts by 1.341 to obtain horsepower.)
[a]based on $\rho_{water} = 1000$ kg/m³ and $g = 9.81$ m/s²
[b]Velocity head changes must be included in Δp.

11. PUMPING EFFICIENCY

$$BHP = \frac{WHP}{\eta_p} \qquad 18.11$$

$$FHP = BHP - WHP \qquad 18.12$$

$$P_{m,hp} = \frac{BHP}{\eta_m} \qquad 18.13$$

$$P_{m,kW} = \frac{0.7457 BHP}{\eta_m} \qquad 18.14$$

$$\eta = \eta_p \eta_m = \frac{WHP}{P_{m,hp}} \qquad 18.15$$

12. COST OF ELECTRICITY

$$\text{cost} = \frac{W_{kW\text{-}hr}(\text{cost per kW-hr})}{\eta_m} \qquad 18.17$$

13. STANDARD MOTOR SIZES AND SPEEDS

- *synchronous speed*

$$n = \frac{120f}{\text{no. of poles}} \qquad 18.19$$

slip (in rpm)

$$= \text{synchronous speed} - \text{actual speed} \qquad 18.20$$

slip (in percent)

$$= \frac{\text{synchronous speed} - \text{actual speed}}{\text{synchronous speed}} \times 100\%$$

$$18.21$$

$$\text{kVA rating} = \frac{\text{motor power in kW}}{\text{power factor}} \qquad 18.22$$

14. PUMP SHAFT LOADING

$$T_{\text{in-lbf}} = \frac{63{,}025 P_{\text{hp}}}{n} \qquad 18.23$$

$$T_{\text{ft-lbf}} = \frac{5252 P_{\text{hp}}}{n} \qquad 18.24$$

$$\text{overhung load} = \frac{2KT}{D_{\text{sheave}}} \qquad 18.27$$

15. SPECIFIC SPEED

$$n_s = \frac{n\sqrt{\dot{V}}}{h_A^{0.75}} \qquad \text{[SI]} \quad 18.28(a)$$

$$n_s = \frac{n\sqrt{Q}}{h_A^{0.75}} \qquad \text{[U.S.]} \quad 18.28(b)$$

17. NET POSITIVE SUCTION HEAD

$$\frac{\text{NPSHR}_2}{\text{NPSHR}_1} = \left(\frac{Q_2}{Q_1}\right)^2 \qquad 18.29$$

$$\text{NPSHA} = h_{\text{atm}} + h_{z(s)} - h_{f(s)} - h_{\text{vp}} - h_{\text{ac}} \qquad 18.32$$

$$\text{NPSHA} = h_{p(s)} + h_{v(s)} - h_{\text{vp}} - h_{\text{ac}} \qquad 18.33$$

$$\text{NPSHA} < \text{NPSHR} \qquad \begin{bmatrix}\text{criterion for}\\ \text{cavitation}\end{bmatrix} \qquad 18.34$$

19. CAVITATION COEFFICIENT

$$\sigma < \sigma_{\text{cr}} \quad \text{[criterion for cavitation]} \qquad 18.35$$

$$\sigma = \frac{2(p - p_{\text{vp}})}{\rho \text{v}^2} = \frac{\text{NPSHA}}{h_A} \qquad \text{[SI]} \quad 18.36(a)$$

$$\sigma = \frac{2g_c(p - p_{\text{vp}})}{\rho \text{v}^2} = \frac{\text{NPSHA}}{h_A} \qquad \text{[U.S.]} \quad 18.36(b)$$

20. SUCTION SPECIFIC SPEED

$$n_{\text{ss}} = \frac{n\sqrt{\dot{V}}}{(\text{NPSHR in m})^{0.75}} \qquad \text{[SI]} \quad 18.37(a)$$

$$n_{\text{ss}} = \frac{n\sqrt{Q}}{(\text{NPSHR in ft})^{0.75}} \qquad \text{[U.S.]} \quad 18.37(b)$$

22. SYSTEM CURVES

$$\frac{h_{f,1}}{h_{f,2}} = \left(\frac{Q_1}{Q_2}\right)^2 \qquad 18.41$$

26. AFFINITY LAWS

- *impeller size is constant, speed varies*

$$\frac{Q_2}{Q_1} = \frac{n_2}{n_1} \qquad 18.42$$

$$\frac{h_2}{h_1} = \left(\frac{n_2}{n_1}\right)^2 = \left(\frac{Q_2}{Q_1}\right)^2 \qquad 18.43$$

$$\frac{P_2}{P_1} = \left(\frac{n_2}{n_1}\right)^3 = \left(\frac{Q_2}{Q_1}\right)^3 \qquad 18.44$$

- *speed is constant, diameter varies*

$$\frac{Q_2}{Q_1} = \frac{D_2}{D_1} \qquad 18.45$$

$$\frac{h_2}{h_1} = \left(\frac{D_2}{D_1}\right)^2 \qquad 18.46$$

$$\frac{P_2}{P_1} = \left(\frac{D_2}{D_1}\right)^3 \qquad 18.47$$

27. PUMP SIMILARITY

- *impeller size varies*

$$\frac{n_1 D_1}{\sqrt{h_1}} = \frac{n_2 D_2}{\sqrt{h_2}} \qquad 18.49$$

$$\frac{Q_1}{D_1^2 \sqrt{h_1}} = \frac{Q_2}{D_2^2 \sqrt{h_2}} \qquad 18.50$$

$$\frac{P_1}{\rho_1 D_1^2 h_1^{1.5}} = \frac{P_2}{\rho_2 D_2^2 h_2^{1.5}} \qquad 18.51$$

$$\frac{Q_1}{n_1 D_1^3} = \frac{Q_2}{n_2 D_2^3} \qquad 18.52$$

$$\frac{P_1}{\rho_1 n_1^3 D_1^5} = \frac{P_2}{\rho_2 n_2^3 D_2^5} \qquad 18.53$$

$$\frac{n_1 \sqrt{Q_1}}{h_1^{0.75}} = \frac{n_2 \sqrt{Q_2}}{h_2^{0.75}} \qquad 18.54$$

29. TURBINE SPECIFIC SPEED

$$n_s = \frac{n\sqrt{P_{\mathrm{kW}}}}{h_t^{1.25}} \qquad \text{[SI]} \quad 18.56(a)$$

$$n_s = \frac{n\sqrt{P_{\mathrm{hp}}}}{h_t^{1.25}} \qquad \text{[U.S.]} \quad 18.56(b)$$

CERM Chapter 19
Open Channel Flow

5. PARAMETERS USED IN OPEN CHANNEL FLOW

- *hydraulic radius*

$$R = \frac{A}{P} \qquad 19.2$$

- *hydraulic depth*

$$D_h = \frac{A}{T} \qquad 19.3$$

- *energy gradient (geometric slope)*

$$S = \frac{dE}{dL} \qquad 19.6$$

$$S_0 = \frac{\Delta z}{L} \approx S \quad \text{[uniform flow]} \qquad 19.7$$

6. GOVERNING EQUATIONS FOR UNIFORM FLOW

$$n = \text{Manning constant}$$

- *Chezy equation*

$$C = \left(\frac{1}{n}\right) R^{1/6} \qquad \text{[SI]} \quad 19.11(a)$$

$$C = \left(\frac{1.49}{n}\right) R^{1/6} \qquad \text{[U.S.]} \quad 19.11(b)$$

- *Chezy-Manning equation*

$$\mathrm{v} = \left(\frac{1}{n}\right) R^{2/3} \sqrt{S} \qquad \text{[SI]} \quad 19.12(a)$$

$$\mathrm{v} = \left(\frac{1.49}{n}\right) R^{2/3} \sqrt{S} \qquad \text{[U.S.]} \quad 19.12(b)$$

$$Q = \mathrm{v}A = \left(\frac{1}{n}\right) A R^{2/3} \sqrt{S}$$
$$= K\sqrt{S} \qquad \text{[SI]} \quad 19.13(a)$$

$$Q = \mathrm{v}A = \left(\frac{1.49}{n}\right) A R^{2/3} \sqrt{S}$$
$$= K\sqrt{S} \qquad \text{[U.S.]} \quad 19.13(b)$$

8. HAZEN-WILLIAMS VELOCITY

$$\mathrm{v} = 0.85\, C R^{0.63} S_0^{0.54} \qquad \text{[SI]} \quad 19.14(a)$$

$$\mathrm{v} = 1.318\, C R^{0.63} S_0^{0.54} \qquad \text{[U.S.]} \quad 19.14(b)$$

9. NORMAL DEPTH

$$d_n = \left(\frac{nQ}{w\sqrt{S}}\right)^{3/5} \quad [w \gg d_n] \qquad \text{[SI]} \quad 19.15(a)$$

$$d_n = 0.788 \left(\frac{nQ}{w\sqrt{S}}\right)^{3/5} \quad [w \gg d_n] \quad \text{[U.S.]} \quad 19.15(b)$$

$$D = d_n = 1.548 \left(\frac{nQ}{\sqrt{S}}\right)^{3/8} \quad \text{[full]} \qquad \text{[SI]} \quad 19.16(a)$$

$$D = d_n = 1.335 \left(\frac{nQ}{\sqrt{S}}\right)^{3/8} \quad \text{[full]} \qquad \text{[U.S.]} \quad 19.16(b)$$

$$D = 2d_n = 2.008 \left(\frac{nQ}{\sqrt{S}}\right)^{3/8} \quad \text{[half full]} \quad \text{[SI]} \quad 19.17(a)$$

$$D = 2d_n = 1.731 \left(\frac{nQ}{\sqrt{S}}\right)^{3/8} \quad \text{[half full]} \quad \text{[U.S.]} \quad 19.17(b)$$

$$R = \frac{wd_n}{w + 2d_n} \qquad 19.18$$

$$A = wd_n \qquad 19.19$$

$$Q = \left(\frac{1}{n}\right) wd_n \left(\frac{wd_n}{w + 2d_n}\right)^{2/3} \sqrt{S} \quad \text{[rectangular]}$$
$$\text{[SI]} \quad 19.20(a)$$

$$Q = \left(\frac{1.49}{n}\right) wd_n \left(\frac{wd_n}{w + 2d_n}\right)^{2/3} \sqrt{S} \quad \text{[rectangular]}$$
$$\text{[U.S.]} \quad 19.20(b)$$

11. SIZING TRAPEZOIDAL AND RECTANGULAR CHANNELS

$$Q = \frac{K' b^{8/3} \sqrt{S_0}}{n} \qquad 19.31$$

$$K' = \left(\frac{\left(1 + m\left(\frac{d}{b}\right)\right)^{5/3}}{\left(1 + 2\left(\frac{d}{b}\right)\sqrt{1 + m^2}\right)^{2/3}} \right) \left(\frac{d}{b}\right)^{5/3} \quad \text{[SI]} \quad 19.32(a)$$

$$K' = \left(\frac{1.49\left(1 + m\left(\frac{d}{b}\right)\right)^{5/3}}{\left(1 + 2\left(\frac{d}{b}\right)\sqrt{1 + m^2}\right)^{2/3}} \right) \left(\frac{d}{b}\right)^{5/3} \quad \text{[U.S.]} \quad 19.32(b)$$

Figure 19.2 *Trapezoidal Cross Section*

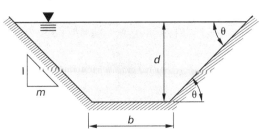

12. MOST EFFICIENT CROSS SECTION

- *rectangle*

$$d = \frac{w}{2} \quad \text{[most efficient rectangle]} \qquad 19.34$$

$$A = dw = \frac{w^2}{2} = 2d^2 \qquad 19.35$$

$$P = d + w + d = 2w = 4d \qquad 19.36$$

$$R = \frac{w}{4} = \frac{d}{2} \qquad 19.37$$

- *trapezoid*

$$d = 2R \quad \text{[most efficient trapezoid]} \qquad 19.38$$

$$b = \frac{2d}{\sqrt{3}} \qquad 19.39$$

$$A = \sqrt{3}d^2 \qquad 19.40$$

$$P = 3b = 2\sqrt{3}d \quad \text{[most efficient trapezoid]} \qquad 19.41$$

$$R = \frac{d}{2} \qquad 19.42$$

14. FLOW MEASUREMENT WITH WEIRS

- *rectangular weirs*

$$Q = \tfrac{2}{3} b \sqrt{2g} \left(\left(H + \frac{v_1^2}{2g} \right)^{3/2} - \left(\frac{v_1^2}{2g} \right)^{3/2} \right) \qquad 19.47$$

$$Q = \tfrac{2}{3} b \sqrt{2g} H^{3/2} \qquad 19.48$$

$$Q = \tfrac{2}{3} C_1 b \sqrt{2g} H^{3/2} \qquad 19.49$$

$$C_1 = \left(0.6035 + 0.0813\left(\frac{H}{Y}\right) + \frac{0.000295}{Y} \right)$$
$$\times \left(1 + \frac{0.00361}{H} \right)^{3/2} \quad \text{[U.S. only]} \qquad 19.50$$

$$C_1 \approx 0.602 + 0.083\left(\frac{H}{Y}\right) \quad \text{[U.S. and SI]} \qquad 19.51$$

$$Q \approx 1.84 b h^{3/2} \qquad \text{[SI]} \qquad 19.52(a)$$

$$Q \approx 3.33 b h^{3/2} \qquad \text{[U.S.]} \qquad 19.52(b)$$

$$b_{\text{effective}} = b_{\text{actual}} - 0.1NH \qquad 19.53$$

17. BROAD-CRESTED WEIRS AND SPILLWAYS

$$Q = \tfrac{2}{3} C_1 b \sqrt{2g} H^{3/2} \qquad 19.59$$

$$Q = C_s b \left(H + \frac{v^2}{2g} \right)^{3/2} \qquad 19.60$$

$$Q = C_s b H^{3/2} \qquad 19.61$$

19. FLOW MEASUREMENT WITH PARSHALL FLUMES

$$Q = K b H_a^n \qquad 19.64$$

$$n = 1.522 b^{0.026} \qquad 19.65$$

21. SPECIFIC ENERGY

$$E = d + \frac{v^2}{2g} \qquad 19.67$$

$$E = d + \frac{Q^2}{2gA^2} \quad \text{[general case]} \qquad 19.68$$

$$E = d + \frac{Q^2}{2g(wd)^2} \quad \text{[rectangular]} \qquad 19.70$$

25. CRITICAL FLOW AND CRITICAL DEPTH IN RECTANGULAR CHANNELS

$$d_c^3 = \frac{Q^2}{gw^2} \quad \text{[rectangular]} \qquad 19.75$$

$$d_c = \tfrac{2}{3}E_c \qquad 19.76$$

$$v_c = \sqrt{gd_c} \qquad 19.77$$

26. CRITICAL FLOW AND CRITICAL DEPTH IN NONRECTANGULAR CHANNELS

$$\frac{Q^2}{g} = \frac{A^3}{T} \quad \text{[nonrectangular]} \qquad 19.78$$

27. FROUDE NUMBER

$$\text{Fr} = \frac{v}{\sqrt{gL}} \qquad 19.79$$

$$\text{Fr} = \frac{\frac{Q}{b}}{\sqrt{gd^3}} \quad \text{[rectangular]} \qquad 19.80$$

$$\text{Fr} = \frac{\frac{Q}{b_{\text{ave}}}}{\sqrt{g\left(\frac{A}{b_{\text{ave}}}\right)^3}} \quad \text{[nonrectangular]} \qquad 19.81$$

33. HYDRAULIC JUMP

$$d_1 = -\tfrac{1}{2}d_2 + \sqrt{\frac{2v_2^2 d_2}{g} + \frac{d_2^2}{4}} \quad \begin{bmatrix}\text{rectangular}\\\text{channels}\end{bmatrix} \qquad 19.91$$

$$d_2 = -\tfrac{1}{2}d_1 + \sqrt{\frac{2v_1^2 d_1}{g} + \frac{d_1^2}{4}} \quad \begin{bmatrix}\text{rectangular}\\\text{channels}\end{bmatrix} \qquad 19.92$$

$$\frac{d_2}{d_1} = \tfrac{1}{2}\left(\sqrt{1 + 8(\text{Fr}_1)^2} - 1\right) \quad \begin{bmatrix}\text{rectangular}\\\text{channels}\end{bmatrix} \qquad 19.93(a)$$

$$\frac{d_1}{d_2} = \tfrac{1}{2}\left(\sqrt{1 + 8(\text{Fr}_2)^2} - 1\right) \quad \begin{bmatrix}\text{rectangular}\\\text{channels}\end{bmatrix} \qquad 19.93(b)$$

$$v_1^2 = \left(\frac{gd_2}{2d_1}\right)(d_1 + d_2) \quad \begin{bmatrix}\text{rectangular}\\\text{channels}\end{bmatrix} \qquad 19.94$$

$$\Delta E = \left(d_1 + \frac{v_1^2}{2g}\right) - \left(d_2 + \frac{v_2^2}{2g}\right) \approx \frac{(d_2 - d_1)^3}{4d_1 d_2} \qquad 19.95$$

38. DETERMINING TYPE OF CULVERT FLOW

(See Fig. 19.25.)

A. Type-1 Flow

$$Q = C_d A_c \sqrt{2g\left(h_1 - z + \frac{\alpha v_1^2}{2g} - d_c - h_{f,1\text{-}2}\right)} \qquad 19.102$$

B. Type-2 Flow

$$Q = C_d A_c \sqrt{2g\left(h_1 + \frac{\alpha v_1^2}{2g} - d_c - h_{f,1\text{-}2} - h_{f,2\text{-}3}\right)} \qquad 19.103$$

C. Type-3 Flow

$$Q = C_d A_3 \sqrt{2g\left(h_1 + \frac{\alpha v_1^2}{2g} - h_3 - h_{f,1\text{-}2} - h_{f,2\text{-}3}\right)} \qquad 19.104$$

D. Type-4 Flow

$$Q = C_d A_o \sqrt{2g\left(\frac{h_1 - h_4}{1 + \dfrac{29 C_d^2 n^2 L}{R^{4/3}}}\right)} \qquad 19.105$$

$$Q = C_d A_o \sqrt{2g(h_1 - h_4)} \qquad 19.106$$

E. Type-5 Flow

$$Q = C_d A_o \sqrt{2g(h_1 - z)} \qquad 19.107$$

F. Type-6 Flow

$$Q = C_d A_o \sqrt{2g(h_1 - h_3 - h_{f,2\text{-}3})} \qquad 19.108$$

Figure 19.25 *Culvert Flow Classifications*

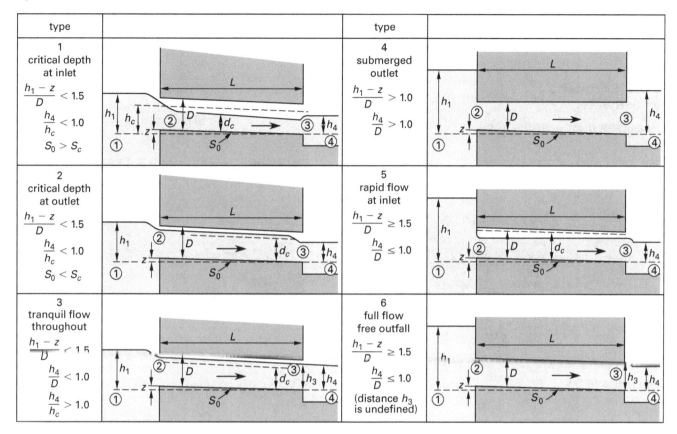

CERM Chapter 20
Meteorology, Climatology, and Hydrology

5. TIME OF CONCENTRATION

$$t_c = t_{\text{sheet}} + t_{\text{shallow}} + t_{\text{channel}} \qquad 20.5$$

$$t_{\text{sheet flow}} = \frac{0.007(nL_o)^{0.8}}{\sqrt{P_2}\,S_{\text{decimal}}^{0.4}} \qquad 20.6$$

$$v_{\text{shallow,ft/sec}} = 16.1345\sqrt{S_{\text{decimal}}} \quad [\text{unpaved}] \qquad 20.7$$

$$v_{\text{shallow,ft/sec}} = 20.3282\sqrt{S_{\text{decimal}}} \quad [\text{paved}] \qquad 20.8$$

$$t_{c,\text{min}} = 1.67 t_{\text{watershed lag time,min}}$$

$$= \frac{(1.67)\left(60\,\frac{\text{min}}{\text{hr}}\right)L_{o,\text{ft}}^{0.8}(S_{\text{in}} + 1)^{0.7}}{1900\sqrt{S_{\text{percent}}}}$$

$$= \frac{(1.67)\left(60\,\frac{\text{min}}{\text{hr}}\right)L_{o,\text{ft}}^{0.8}\left(\frac{1000}{\text{CN}} - 9\right)^{0.7}}{1900\sqrt{S_{\text{percent}}}} \qquad 20.11$$

6. RAINFALL INTENSITY

$$I = \frac{K}{t_c + b} \qquad 20.14$$

7. FLOODS

$$p\{F \text{ event in } n \text{ years}\} = 1 - \left(1 - \frac{1}{F}\right)^n \qquad 20.20$$

10. UNIT HYDROGRAPH

$$V = A_d P_{\text{ave,excess}} \qquad 20.21$$

15. PEAK RUNOFF FROM THE RATIONAL METHOD

$$Q_p = CIA_d \qquad 20.36$$

20. RESERVOIR SIZING: RESERVOIR ROUTING

$$V_{n+1} = V_n + (\text{inflow})_n - (\text{discharge})_n$$
$$- (\text{seepage})_n - (\text{evaporation})_n \qquad 20.47$$

CERM Chapter 21
Groundwater

2. AQUIFER CHARACTERISTICS

$$w = \frac{m_w}{m_s} = \frac{m_t - m_s}{m_s} \qquad 21.2$$

$$n = \frac{V_v}{V_t} = \frac{V_t - V_s}{V_t} \qquad 21.3$$

$$e = \frac{V_v}{V_s} = \frac{V_t - V_s}{V_s} \qquad 21.4$$

$$e = \frac{n}{1-n} \qquad 21.5$$

$$i = \frac{\Delta H}{L} \qquad 21.6$$

3. PERMEABILITY

$$K = \frac{kg\rho}{\mu} \qquad [\text{SI}] \quad 21.7(a)$$

$$K = \frac{k\gamma}{\mu} \qquad [\text{U.S.}] \quad 21.7(b)$$

$$K = CD_{\text{mean}}^2 \qquad 21.8$$

$$K_{\text{cm/s}} \approx CD_{10,\text{mm}}^2 \quad [0.1 \text{ mm} \le D_{10,\text{mm}} \le 3.0 \text{ mm}]$$
$$21.9$$

(fine sand $0.4 \le C \le 1.5$ coarse sand)

4. DARCY'S LAW

$$Q = -KiA_{\text{gross}} = -v_e A_{\text{gross}} \qquad 21.10$$

$$q = \frac{Q}{A_{\text{gross}}} = Ki = v_e \qquad 21.11$$

$$\text{Re} = \frac{\rho q D_{\text{mean}}}{\mu} = \frac{q D_{\text{mean}}}{\nu} \qquad 21.12$$

5. TRANSMISSIVITY

$$T = KY \qquad 21.13$$
$$Q = bTi \qquad 21.14$$

6. SPECIFIC YIELD, RETENTION, AND CAPACITY

$$S_y = \frac{V_{\text{yielded}}}{V_{\text{total}}} \qquad 21.15$$

$$S_r = \frac{V_{\text{retained}}}{V_{\text{total}}} = n - S_y \qquad 21.16$$

$$\text{specific capacity} = \frac{Q}{s} \qquad 21.17$$

7. DISCHARGE VELOCITY AND SEEPAGE VELOCITY

$$v_{\text{pore}} = \frac{Q}{A_{\text{net}}} = \frac{Q}{nA_{\text{gross}}} = \frac{Q}{nbY} = \frac{Ki}{n} \qquad 21.18$$

$$v_e = nv_{\text{pore}} = \frac{Q}{A_{\text{gross}}} = \frac{Q}{bY} = Ki \qquad 21.19$$

10. DESIGN OF GRAVEL SCREENS AND POROUS FILTERS

$$D_{\text{opening,filter}} \le D_{85,\text{soil}} \quad [\text{screen filters}] \qquad 21.21$$

$$[\text{filtering criterion}] \; D_{15,\text{filter}} \le 5D_{85,\text{soil}} \quad [\text{filter beds}]$$
$$21.22$$

$$[\text{permeability criterion}] \; D_{15,\text{filter}} \le 5D_{15,\text{soil}} \quad [\text{filter beds}]$$
$$21.23$$

$$C_u = \frac{D_{60}}{D_{10}} \qquad 21.24$$

11. WELL DRAWDOWN IN AQUIFERS

- *Dupuit equation*

$$Q = \frac{\pi K(y_1^2 - y_2^2)}{\ln \dfrac{r_1}{r_2}} \qquad 21.25$$

- $s \ll Y$

$$Q = \frac{2\pi T(s_2 - s_1)}{\ln \dfrac{r_1}{r_2}} \qquad 21.26$$

- *Thiem equation (artesian well)*

$$Q = \frac{2\pi KY(y_1 - y_2)}{\ln \dfrac{r_1}{r_2}} \qquad 21.27$$

12. UNSTEADY FLOW

- *Theis equation*

$$s_{r,t} = \left(\frac{Q}{4\pi KY}\right) W(u)$$

$$= \left(\frac{Q}{4\pi T}\right) W(u) \quad \begin{bmatrix} \text{consistent} \\ \text{units} \end{bmatrix} \qquad 21.28$$

$$s_1 - s_2 = y_2 - y_1 = \left(\frac{Q}{4\pi KY}\right)\left(W(u_1) - W(u_2) \right) \qquad 21.31$$

15. SEEPAGE FROM FLOW NETS

$$Q = KH\left(\frac{N_f}{N_p}\right) \quad \text{[per unit width]} \qquad 21.33$$

$$H = H_1 - H_2 \qquad 21.34$$

16. HYDROSTATIC PRESSURE ALONG FLOW PATH

$$p_u = \left(\frac{j}{N_p}\right) Hg\rho_w \qquad \text{[SI]} \quad 21.35(a)$$

$$p_u = \left(\frac{j}{N_p}\right) H\gamma_w \qquad \text{[U.S.]} \quad 21.35(b)$$

$$p_u = \left(\left(\frac{j}{N_p}\right) H + z\right) g\rho_w \qquad \text{[SI]} \quad 21.36(a)$$

$$p_u = \left(\left(\frac{j}{N_p}\right) H + z\right) \gamma_w \qquad \text{[U.S.]} \quad 21.36(b)$$

$$U = Np_u A \qquad 21.37$$

$$(\text{FS})_{\text{heave}} = \frac{\text{downward pressure}}{\text{uplift pressure}} \qquad 21.38$$

17. INFILTRATION

$$f_t = f_c + (f_0 - f_c)e^{-kt} \qquad 21.39$$

$$F_t = f_c t + \left(\frac{f_0 - f_c}{k}\right)(1 - e^{-kt}) \qquad 21.42$$

$$\bar{f} = \frac{F_t}{t_{\text{infiltration}}} \qquad 21.43$$

$$S_{\max} = \frac{f_0 - f_c}{k} \qquad 21.45$$

CERM Chapter 22
Inorganic Chemistry

9. EQUIVALENT WEIGHT

$$\text{EW} = \frac{\text{MW}}{\Delta \text{ oxidation number}} \qquad 22.2$$

10. GRAVIMETRIC FRACTION

$$x_i = \frac{m_i}{m_1 + m_2 + \cdots + m_i + \cdots + m_n} = \frac{m_i}{m_t} \qquad 22.3$$

18. SOLUTIONS OF GASES IN LIQUIDS

- *Henry's law*

$$p_i = H_i x_i \qquad 22.7$$

21. UNITS OF CONCENTRATION

F— *formality:* The number of gram formula weights (i.e., molecular weights in grams) per liter of solution.

m— *molality:* The number of gram-moles of solute per 1000 grams of solvent. A "molal" solution contains 1 gram-mole per 1000 grams of solvent.

M— *molarity:* The number of gram-moles of solute per liter of solution. A "molar" (i.e., 1 M) solution contains 1 gram-mole per liter of solution. Molarity is related to normality as shown in Eq. 22.8.

$$N = M \times \Delta \text{ oxidation number} \qquad 22.8$$

N— *normality:* The number of gram equivalent weights of solute per liter of solution. A solution is "normal" (i.e., 1 N) if there is exactly one gram equivalent weight per liter of solution.

x— *mole fraction:* The number of moles of solute divided by the number of moles of solvent and all solutes.

meq/L— *milligram equivalent weights of solute per liter of solution:* calculated by multiplying normality by 1000 or dividing concentration in mg/L by equivalent weight.

mg/L— *milligrams per liter:* The number of milligrams of solute per liter of solution. Same as ppm for solutions of water.

ppm— *parts per million:* The number of pounds (or grams) of solute per million pounds (or grams) of solution. Same as mg/L for solutions of water.

ppb— *parts per billion:* The number of pounds (or grams) of solute per billion (10^9) pounds (or grams) of solution. Same as μg/L for solutions of water.

22. pH AND pOH

$$\text{pH} = -\log_{10}[\text{H}^+] = \log_{10}\frac{1}{[\text{H}^+]} \qquad 22.9$$

$$\text{pOH} = -\log_{10}[\text{OH}^-] = \log_{10}\frac{1}{[\text{OH}^-]} \qquad 22.10$$

$$[\text{ion}] = XM \qquad 22.11$$

$$\text{pH} + \text{pOH} = 14 \qquad 22.12$$

24. NEUTRALIZATION

$$V_b N_b = V_a N_a \qquad 22.13$$

$$V_b M_b \Delta_{b,\text{charge}} = V_a M_a \Delta_{a,\text{charge}} \qquad 22.14$$

29. REVERSIBLE REACTION KINETICS

$$a\text{A} + b\text{B} \rightleftharpoons c\text{C} + d\text{D} \qquad 22.16$$

$$\text{v}_{\text{forward}} = k_{\text{forward}}[\text{A}]^a[\text{B}]^b \qquad 22.17$$

$$\text{v}_{\text{reverse}} = k_{\text{reverse}}[\text{C}]^c[\text{D}]^d \qquad 22.18$$

30. EQUILIBRIUM CONSTANT

$$K = \frac{[\text{C}]^c[\text{D}]^d}{[\text{A}]^a[\text{B}]^b} = \frac{k_{\text{forward}}}{k_{\text{reverse}}} \qquad 22.20$$

31. IONIZATION CONSTANT

$$K_{\text{ionization}} = \frac{MX^2}{1-X} \quad [K_a \text{ or } K_b] \qquad 22.25$$

35. ENTHALPY OF REACTION

$$\Delta H_r = \sum_{\text{products}} \Delta H_f - \sum_{\text{reactants}} \Delta H_f \qquad 22.27$$

(See the "Atomic Numbers and Weights of the Elements" table.)

36. GALVANIC ACTION

Table 22.6 Galvanic Series in Seawater (top to bottom anodic (sacrificial, active) to cathodic (noble, passive))

magnesium
zinc
Alclad 3S
cadmium
2024 aluminum alloy
low-carbon steel
cast iron
stainless steels (active)
   no. 410
   no. 430
   no. 404
   no. 316
Hastelloy A
lead
lead-tin alloys
tin
nickel
brass (copper-zinc)
copper
bronze (copper-tin)
90/10 copper-nickel
70/30 copper-nickel
Inconel
silver solder
silver
stainless steels (passive)
Monel metal
Hastelloy C
titanium
graphite
gold

CERM Chapter 24
Combustion and Incineration

6. MOISTURE

$$G_{\text{H,combined}} = \frac{G_\text{O}}{8} \qquad 24.1$$

$$G_{\text{H,available}} = G_{\text{H,total}} - \frac{G_\text{O}}{8} \qquad 24.2$$

Atomic Numbers and Weights of the Elements

name	symbol	atomic number	atomic weight	name	symbol	atomic number	atomic weight
actinium	Ac	89	–	meitnerium	Mt	109	–
aluminum	Al	13	26.9815	mendelevium	Md	101	–
americium	Am	95	–	mercury	Hg	80	200.59
antimony	Sb	51	121.760	molybdenum	Mo	42	95.96
argon	Ar	18	39.948	neodymium	Nd	60	144.242
arsenic	As	33	74.9216	neon	Ne	10	20.1797
astatine	At	85	–	neptunium	Np	93	237.048
barium	Ba	56	137.327	nickel	Ni	28	58.693
berkelium	Bk	97	–	niobium	Nb	41	92.906
beryllium	Be	4	9.0122	nitrogen	N	7	14.0067
bismuth	Bi	83	208.980	nobelium	No	102	–
bohrium	Bh	107	–	osmium	Os	76	190.23
boron	B	5	10.811	oxygen	O	8	15.9994
bromine	Br	35	79.904	palladium	Pd	46	106.42
cadmium	Cd	48	112.411	phosphorus	P	15	30.9738
calcium	Ca	20	40.078	platinum	Pt	78	195.084
californium	Cf	98	–	plutonium	Pu	94	–
carbon	C	6	12.0107	polonium	Po	84	–
cerium	Ce	58	140.116	potassium	K	19	39.0983
cesium	Cs	55	132.9054	praseodymium	Pr	59	140.9077
chlorine	Cl	17	35.453	promethium	Pm	61	–
chromium	Cr	24	51.996	protactinium	Pa	91	231.0359
cobalt	Co	27	58.9332	radium	Ra	88	–
copernicium	Cn	112	–	radon	Rn	86	226.025
copper	Cu	29	63.546	rhenium	Re	75	186.207
curium	Cm	96	–	rhodium	Rh	45	102.9055
darmstadtium	Ds	110	–	roentgenium	Rg	111	–
dubnium	Db	105	–	rubidium	Rb	37	85.4678
dysprosium	Dy	66	162.50	ruthenium	Ru	44	101.07
einsteinium	Es	99	–	rutherfordium	Rf	104	–
erbium	Er	68	167.259	samarium	Sm	62	150.36
europium	Eu	63	151.964	scandium	Sc	21	44.956
fermium	Fm	100	–	seaborgium	Sg	106	–
fluorine	F	9	18.9984	selenium	Se	34	78.96
francium	Fr	87	–	silicon	Si	14	28.0855
gadolinium	Gd	64	157.25	silver	Ag	47	107.868
gallium	Ga	31	69.723	sodium	Na	11	22.9898
germanium	Ge	32	72.64	strontium	Sr	38	87.62
gold	Au	79	196.9666	sulfur	S	16	32.065
hafnium	Hf	72	178.49	tantalum	Ta	73	180.94788
hassium	Hs	108	–	technetium	Tc	43	–
helium	He	2	4.0026	tellurium	Te	52	127.60
holmium	Ho	67	164.930	terbium	Tb	65	158.925
hydrogen	H	1	1.00794	thallium	Tl	81	204.383
indium	In	49	114.818	thorium	Th	90	232.038
iodine	I	53	126.90447	thulium	Tm	69	168.934
iridium	Ir	77	192.217	tin	Sn	50	118.710
iron	Fe	26	55.845	titanium	Ti	22	47.867
krypton	Kr	36	83.798	tungsten	W	74	183.84
lanthanum	La	57	138.9055	uranium	U	92	238.0289
lawrencium	Lr	103	–	vanadium	V	23	50.942
lead	Pb	82	207.2	xenon	Xe	54	131.293
lithium	Li	3	6.941	ytterbium	Yb	70	173.054
lutetium	Lu	71	174.9668	yttrium	Y	39	88.906
magnesium	Mg	12	24.305	zinc	Zn	30	65.38
manganese	Mn	25	54.9380	zirconium	Zr	40	91.224

24. ATMOSPHERIC AIR

Table 24.6 *Composition of Dry Air*[a]

component	percent by weight	percent by volume
oxygen	23.15	20.95
nitrogen/inerts	76.85	79.05
ratio of nitrogen to oxygen	3.320	3.773[b]
ratio of air to oxygen	4.320	4.773

[a]Inert gases and CO_2 are included as N_2.
[b]The value is also reported by various sources as 3.76, 3.78, and 3.784.

25. COMBUSTION REACTIONS

Table 24.7 *Ideal Combustion Reactions*

fuel	formula	reaction equation (excluding nitrogen)
carbon (to CO)	C	$2C + O_2 \rightarrow 2CO$
carbon (to CO_2)	C	$C + O_2 \rightarrow CO_2$
sulfur (to SO_2)	S	$S + O_2 \rightarrow SO_2$
sulfur (to SO_3)	S	$2S + 3O_2 \rightarrow 2SO_3$
carbon monoxide	CO	$2CO + O_2 \rightarrow 2CO_2$
methane	CH_4	$CH_4 + 2O_2 \rightarrow CO_2 + 2H_2O$
acetylene	C_2H_2	$2C_2H_2 + 5O_2 \rightarrow 4CO_2 + 2H_2O$
ethylene	C_2H_4	$C_2H_4 + 3O_2 \rightarrow 2CO_2 + 2H_2O$
ethane	C_2H_6	$2C_2H_6 + 7O_2 \rightarrow 4CO_2 + 6H_2O$
hydrogen	H_2	$2H_2 + O_2 \rightarrow 2H_2O$
hydrogen sulfide	H_2S	$2H_2S + 3O_2 \rightarrow 2H_2O + 2SO_2$
propane	C_3H_8	$C_3H_8 + 5O_2 \rightarrow 3CO_2 + 4H_2O$
n-butane	C_4H_{10}	$2C_4H_{10} + 13O_2 \rightarrow 8CO_2 + 10H_2O$
octane	C_8H_{18}	$2C_8H_{18} + 25O_2 \rightarrow 16CO_2 + 18H_2O$
olefin series	C_nH_{2n}	$2C_nH_{2n} + 3nO_2 \rightarrow 2nCO_2 + 2nH_2O$
paraffin series	C_nH_{2n+2}	$2C_nH_{2n+2} + (3n+1)O_2 \rightarrow 2nCO_2 + (2n+2)H_2O$

(Multiply oxygen volumes by 3.773 to get nitrogen volumes.)

27. STOICHIOMETRIC AIR

$$R_{a/f,\text{ideal}} = \frac{m_{\text{air,ideal}}}{m_{\text{fuel}}} \quad 24.5$$

$$R_{a/f,\text{ideal}} = 34.5\left(\frac{G_C}{3} + G_H - \frac{G_O}{8} + \frac{G_S}{8}\right)$$
[solid and liquid fuels] 24.6

$$R_{a/f,\text{ideal}} = \sum G_i J_i \quad \text{[gaseous fuels]} \quad 24.7$$

$$\text{volumetric air/fuel ratio} = \sum B_i K_i \quad \text{[gaseous fuels]} \quad 24.8$$

31. ACTUAL AND EXCESS AIR

$$R_{a/f,\text{actual}} = \frac{m_{\text{air,actual}}}{m_{\text{fuel}}}$$
$$= \frac{3.04 B_{N_2}\left(G_C + \frac{G_S}{1.833}\right)}{B_{CO_2} + B_{CO}} \quad 24.9$$

34. DEW POINT OF FLUE GAS MOISTURE

$$\frac{\text{water vapor}}{\text{partial pressure}} = \frac{(\text{water vapor mole fraction})}{\times (\text{flue gas pressure})} \quad 24.13$$

35. HEAT OF COMBUSTION

$$\text{HHV} = \text{LHV} + m_{\text{water}} h_{fg} \quad 24.14$$

$$G_{H,\text{available}} = G_{H,\text{total}} - \frac{G_O}{8} \quad 24.15$$

$$\text{HHV}_{\text{MJ/kg}} = 32.78 G_C + 141.8\left(G_H - \frac{G_O}{8}\right) + 9.264 G_S$$
[SI] 24.16(a)

$$\text{HHV}_{\text{Btu/lbm}} = 14{,}093 G_C + 60{,}958\left(G_H - \frac{G_O}{8}\right) + 3983 G_S$$
[U.S.] 24.16(b)

$$\text{HHV}_{\text{gasoline,MJ/kg}} = 42.61 + 0.093(°\text{Be} - 10)$$
[SI] 24.17(a)

$$\text{HHV}_{\text{gasoline,Btu/lbm}} = 18{,}320 + 40(°\text{Be} - 10)$$
[U.S.] 24.17(b)

$$\text{HHV}_{\text{fuel oil,MJ/kg}} = 51.92 - 8.792(\text{SG})^2 \quad \text{[SI]} \quad 24.18(a)$$

$$\text{HHV}_{\text{fuel oil,Btu/lbm}} = 22{,}320 - 3780(\text{SG})^2 \quad \text{[U.S.]} \quad 24.18(b)$$

36. MAXIMUM THEORETICAL COMBUSTION (FLAME) TEMPERATURE

$$T_{\max} = T_i + \frac{\text{LHV}}{m_{\text{products}} c_{p,\text{mean}}} \qquad 24.19$$

37. COMBUSTION LOSSES

$$q_1 = m_{\text{flue gas}} c_p \left(T_{\text{flue gas}} - T_{\text{incoming air}} \right) \qquad 24.20$$

$$q_2 = m_{\text{vapor}} \left(h_g - h_f \right) = 8.94 G_{\text{H}} \left(h_g - h_f \right) \qquad 24.21$$

$$q_3 = m_{\text{atmospheric water vapor}} \left(h_g - h'_g \right)$$
$$= \omega m_{\text{combustion air}} \left(h_g - h'_g \right) \qquad 24.22$$

$$q_4 = \frac{2.334 \text{HHV}_{\text{CO}} G_{\text{C}} B_{\text{CO}}}{B_{\text{CO}_2} + B_{\text{CO}}} \qquad 24.23$$

$$q_5 = \text{HHV}_{\text{C}} m_{\text{ash}} G_{\text{C,ash}} \qquad 24.24$$

38. COMBUSTION EFFICIENCY

$$\eta = \frac{\text{useful heat extracted}}{\text{heating value}}$$
$$= \frac{m_{\text{steam}} \left(h_{\text{steam}} - h_{\text{feedwater}} \right)}{\text{HHV}} \qquad 24.25$$

$$\eta = \frac{\text{HHV} - q_1 - q_2 - q_3 - q_4 - q_5 - \text{radiation}}{\text{HHV}}$$
$$= \frac{\text{LHV} - q_1 - q_4 - q_5 - \text{radiation}}{\text{HHV}} \qquad 24.26$$

CERM Chapter 25
Water Supply Quality and Testing

2. ACIDITY

$$A_{\text{mg/L as CaCO}_3} = \frac{V_{\text{titrant,mL}} N_{\text{titrant}} (50\,000)}{V_{\text{sample,mL}}} \qquad 25.4$$

3. ALKALINITY

$$M_{\text{mg/L as CaCO}_3} = \frac{V_{\text{titrant,mL}} N_{\text{titrant}} (50\,000)}{V_{\text{sample,mL}}} \qquad 25.5$$

6. HARDNESS AND ALKALINITY

(See Fig. 25.1.)

24. SALINITY IN IRRIGATION WATER

$$\text{SSP} = \frac{[\text{Na}^+] \times 100\%}{[\text{Ca}^{++}] + [\text{Mg}^{++}] + [\text{Na}^+] + [\text{K}^+]} \qquad 25.20$$

$$\text{RSC} = \left([\text{HCO}_3^-] + [\text{CO}_3^{--}] \right) - \left([\text{Ca}^{++}] + [\text{Mg}^{++}] \right) \qquad 25.21$$

$$\text{SAR} = \frac{[\text{Na}^+]}{\sqrt{\dfrac{[\text{Ca}^{++}] + [\text{Mg}^{++}]}{2}}} \qquad 25.22$$

Conversions from mg/L as a Substance to mg/L as CaCo₃

Multiply the concentration (in mg/L) of the substance by the corresponding factors to obtain the equivalent concentration in mg/L as $CaCO_3$. For example, 70 mg/L of Mg^{++} would be (70 mg/L)(4.10) = 287 mg/L as $CaCO_3$.

substance	factor	substance	factor
Al^{+++}	5.56	HCO_3^-	0.82
$Al_2(SO_4)_3$	0.88*	K^+	1.28
$AlCl_3$	1.13	KCl	0.67
$Al(OH)_3$	1.92	K_2CO_3	0.72
Ba^{++}	0.73	Mg^{++}	4.10
$Ba(OH)_2$	0.59	$MgCl_2$	1.05
$BaSO_4$	0.43	$MgCO_3$	1.19
Ca^{++}	2.50	$Mg(HCO_3)_2$	0.68
$CaCl_2$	0.90	MgO	2.48
$CaCO_3$	1.00	$Mg(OH)_2$	1.71
$Ca(HCO_3)_2$	0.62	$Mg(NO_3)_2$	0.67
CaO	1.79	$MgSO_4$	0.83
$Ca(OH)_2$	1.35	Mn^{++}	1.82
$CaSO_4$	0.74*	Na^+	2.18
Cl^-	1.41	NaCl	0.85
CO_2	2.27	Na_2CO_3	0.94
CO_3^{--}	1.67	$NaHCO_3$	0.60
Cu^{++}	1.57	$NaNO_3$	0.59
Cu^{+++}	2.36	NaOH	1.25
$CuSO_4$	0.63	Na_2SO_4	0.70*
F^-	2.66	NH_3	2.94
Fe^{++}	1.79	NH_4^+	2.78
Fe^{+++}	2.69	NH_4OH	1.43
$Fe(OH)_3$	1.41	$(NH_4)_2SO_4$	0.76
$FeSO_4$	0.66*	NO_3^-	0.81
$Fe_2(SO_4)_3$	0.75	OH^-	2.94
$FeCl_3$	0.93	PO_4^{---}	1.58
H^+	50.0	SO_4^{--}	1.04
		Zn^{++}	1.54

*anhydrous

Figure 25.1 *Hardness and Alkalinity*[a,b]

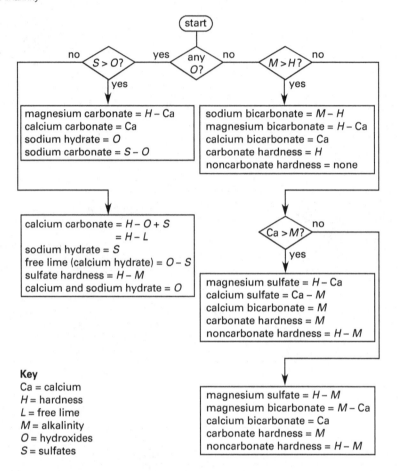

Key
Ca = calcium
H = hardness
L = free lime
M = alkalinity
O = hydroxides
S = sulfates

[a] All concentrations are expressed as $CaCO_3$.
[b] Not for use when other ionic species are present in significant quantities.

CERM Chapter 26
Water Supply Treatment and Distribution

10. AERATION

$$P_{kW} = \frac{Q_{L/s}h_m}{100} \qquad \text{[SI]} \quad 26.1(a)$$

$$P_{hp} = \frac{Q_{cfm}h_{ft}}{528} \qquad \text{[U.S.]} \quad 26.1(b)$$

11. SEDIMENTATION PHYSICS

$$Re = \frac{v_s D}{\nu} \qquad 26.2$$

- Equation 26.3 calculates the settling velocity; the second form shown is known as Stokes' law.

$$v_{s,m/s} = \frac{(\rho_{particle} - \rho_{water})D_m^2 g}{18\mu}$$

$$= \frac{(SG_{particle} - 1)D_m^2 g}{18\nu} \qquad \text{[SI]} \quad 26.3(a)$$

$$v_{s,ft/sec} = \frac{(\rho_{particle} - \rho_{water})D_{ft}^2 g}{18\mu g_c}$$

$$= \frac{(SG_{particle} - 1)D_{ft}^2 g}{18\nu} \qquad \text{[U.S.]} \quad 26.3(b)$$

$$v_s = \sqrt{\frac{4gD(SG_{particle} - 1)}{3C_D}} \qquad 26.4$$

12. SEDIMENTATION TANKS

$$t_{\text{settling}} = \frac{h}{v_s} \qquad 26.5$$

$$t_d = \frac{V_{\text{tank}}}{Q} = \frac{Ah}{Q} \qquad 26.6$$

$$v^* = \frac{Q_{\text{filter}}}{A_{\text{surface}}} = \frac{Q_{\text{filter}}}{bL} \qquad 26.7$$

17. DOSES OF COAGULANTS AND OTHER COMPOUNDS

$$F_{\text{kg/d}} = \frac{D_{\text{mg/L}} Q_{\text{ML/d}}}{PG} \qquad \text{[SI]} \quad 26.16(a)$$

$$F_{\text{lbm/day}} = \frac{D_{\text{mg/L}} Q_{\text{MGD}} \left(8.345 \, \frac{\text{lbm-L}}{\text{mg-MG}} \right)}{PG}$$
$$\text{[U.S.]} \quad 26.16(b)$$

18. MIXERS AND MIXING KINETICS

$$V = tQ \qquad 26.17$$

$$t_{\text{complete}} = \frac{V}{Q} = \frac{1}{K} \left(\frac{C_i}{C_o} - 1 \right) \qquad 26.18$$

$$t_{\text{plug flow}} = \frac{V}{Q} = \frac{L}{v_f}$$
$$= \frac{1}{K} \ln \frac{C_i}{C_o} \qquad 26.19$$

19. MIXING PHYSICS

$$F_D = \frac{C_D A \rho v_{\text{mixing}}^2}{2} \qquad \text{[SI]} \quad 26.20(a)$$

$$F_D = \frac{C_D A \rho v_{\text{mixing}}^2}{2 g_c} = \frac{C_D A \gamma v_{\text{mixing}}^2}{2g} \qquad \text{[U.S.]} \quad 26.20(b)$$

$$v_{\text{paddle,ft/sec}} = \frac{2 \pi R n_{\text{rpm}}}{60 \, \frac{\text{sec}}{\text{min}}} \qquad 26.21$$

$$v_{\text{mixing}} = v_{\text{paddle}} - v_{\text{water}} \qquad 26.22$$

$$P_{\text{kW}} = \frac{F_D v_{\text{mixing}}}{1000 \, \frac{\text{W}}{\text{kW}}}$$
$$= \frac{C_D A \rho v_{\text{mixing}}^3}{2 \left(1000 \, \frac{\text{W}}{\text{kW}} \right)} \qquad \text{[SI]} \quad 26.23(a)$$

$$P_{\text{hp}} = \frac{F_D v_{\text{mixing}}}{550 \, \frac{\text{ft-lbf}}{\text{hp-sec}}} = \frac{C_D A \rho v_{\text{mixing}}^3}{2 g_c \left(550 \, \frac{\text{ft-lbf}}{\text{hp-sec}} \right)}$$
$$= \frac{C_D A \gamma v_{\text{mixing}}^3}{2g \left(550 \, \frac{\text{ft-lbf}}{\text{hp-sec}} \right)}$$
$$\text{[U.S.]} \quad 26.23(b)$$

$$G = \sqrt{\frac{P}{\mu V_{\text{tank}}}} \qquad 26.24$$

$$P = \mu G^2 V_{\text{tank}} \qquad 26.25$$

$$Gt_d = \frac{V_{\text{tank}}}{Q} \sqrt{\frac{P}{\mu V_{\text{tank}}}}$$
$$= \frac{1}{Q} \sqrt{\frac{P V_{\text{tank}}}{\mu}} \qquad 26.26$$

20. IMPELLER CHARACTERISTICS

$$\text{Re} = \frac{D^2 n \rho}{\mu} \qquad \text{[SI]} \quad 26.27(a)$$

$$\text{Re} = \frac{D^2 n \rho}{g_c \mu} \qquad \text{[U.S.]} \quad 26.27(b)$$

- *geometrically similar and turbulent flow*

$$P = N_P n^3 D^5 \rho \qquad \text{[SI]} \quad 26.28(a)$$

$$P = \frac{N_P n^3 D^5 \rho}{g_c} \qquad \text{[U.S.]} \quad 26.28(b)$$

$$P = \rho g Q h_v \qquad \text{[SI]} \quad 26.29(a)$$

$$P = \frac{\rho g Q h_v}{g_c} \qquad \text{[U.S.]} \quad 26.29(b)$$

$$Q = N_Q n D^3 \qquad 26.30$$

$$h_v = \frac{N_P n^2 D^2}{N_Q g} \qquad 26.31$$

23. SLUDGE QUANTITIES

$$V_{\text{sludge}} = \frac{m_{\text{sludge}}}{\rho_{\text{water}} (\text{SG})_{\text{sludge}} G} \qquad 26.33$$

$$m_{\text{sludge,kg/d}} =$$

$$\left(86\,400\,\tfrac{\text{s}}{\text{d}}\right) Q_{\text{m}^3/\text{s}}\left(1000\,\tfrac{\text{L}}{\text{m}^3}\right)$$
$$\frac{\times \left(0.46 D_{\text{alum,mg/L}} + \Delta\text{TSS}_{\text{mg/L}} + M_{\text{mg/L}}\right)}{10^6\,\tfrac{\text{mg}}{\text{kg}}}$$

$$[\text{SI}] \quad 26.34(a)$$

$$m_{\text{sludge,lbm/day}} =$$

$$\left(8.345\,\tfrac{\text{lbm-L}}{\text{mg-MG}}\right) Q_{\text{MGD}}$$
$$\times \left(0.46 D_{\text{alum,mg/L}} + \Delta\text{TSS}_{\text{mg/L}} + M_{\text{mg/L}}\right)$$

$$[\text{U.S.}] \quad 26.34(b)$$

$$V_{2,\text{sludge}} \approx V_{1,\text{sludge}} \frac{G_1}{G_2} \qquad 26.35$$

24. FILTRATION

$$\text{loading rate} = \frac{Q}{A} \qquad 26.36$$

25. FILTER BACKWASHING

$$V = A_{\text{filter}}(\text{rate of rise})t_{\text{backwash}} \qquad 26.37$$

28. FLUORIDATION

$$F_{\text{kg/d}} = \frac{C_{\text{mg/L}} Q_{\text{L/d}}}{PG} \qquad [\text{SI}] \quad 26.38(a)$$

$$F_{\text{lbm/day}} = \frac{C_{\text{mg/L}} Q_{\text{MGD}}\left(8.345\,\tfrac{\text{lbm-L}}{\text{mg-MG}}\right)}{PG}$$

$$[\text{U.S.}] \quad 26.38(b)$$

30. TASTE AND ODOR CONTROL

$$\text{TON} = \frac{V_{\text{raw sample}} + V_{\text{dilution water}}}{V_{\text{raw sample}}} \qquad 26.39$$

33. WATER SOFTENING BY ION EXCHANGE

$$V_{\text{water}} = \frac{(\text{specific working capacity})\,V_{\text{exchange material}}}{\text{hardness}}$$

$$26.48$$

34. REGENERATION OF ION EXCHANGE RESINS

$$m_{\text{salt}} = (\text{specific working capacity})\,V_{\text{exchange material}}$$
$$\times (\text{salt requirement})$$

$$26.49$$

35. STABILIZATION AND SCALING POTENTIAL

$$\text{LSI} = \text{pH} - \text{pH}_{\text{sat}} \qquad 26.50$$

$$\text{RSI} = 2\text{pH}_{\text{sat}} - \text{pH} = \text{pH} - 2(\text{LSI}) \qquad 26.51$$

$$\text{PSI} = 2\text{pH}_{\text{sat}} - \text{pH}_{\text{eq}} \qquad 26.52$$

$$\text{pH}_{\text{eq}} = 4.54 + 1.465\log_{10}[M] \qquad 26.53$$

$$\text{pH}_{\text{sat}} = (\text{p}K_a - \text{p}K_{\text{sp}}) + \text{pCa} + \text{pM}$$
$$= -\log_{10}\left(\frac{K_a[\text{Ca}^{++}][M]}{K_{\text{sp}}}\right) \qquad 26.54$$

$$M = [\text{HCO}_3^-] + 2[\text{CO}_3^{--}] + [\text{OH}^-] \qquad 26.55$$

$$\text{AI} = \text{pH} + \log_{10}[M] + \log_{10}[\text{Ca}^{++}] \qquad 26.61$$

43. WATER DEMAND

$$Q_{\text{instantaneous}} = M(\text{AADF}) \qquad 26.66$$

44. FIRE FIGHTING DEMAND

$$\text{NFF} = CO(1 + X + P) \qquad 26.67$$

$$C = 18F\sqrt{A} \qquad 26.68$$

45. OTHER FORMULAS FOR FIRE FIGHTING NEEDS

$$Q_{\text{gpm}} = 1020\sqrt{P}(1 - 0.01\sqrt{P}) \qquad 26.69$$

Environmental

CERM Chapter 28
Wastewater Quantity and Quality

4. WASTEWATER QUANTITY

- TSS equation (P in thousands):

$$\frac{Q_{\text{peak}}}{Q_{\text{ave}}} = \frac{18 + \sqrt{P}}{4 + \sqrt{P}} \qquad 28.1$$

- Harmon's equation:

$$\frac{Q_{\text{peak}}}{Q_{\text{ave}}} = \frac{1 + 14}{4 + \sqrt{P}} \geq 2.5 \qquad 28.2$$

5. TREATMENT PLANT LOADING

$$P_{e,1000\text{s}} = \frac{\text{BOD}_{\text{mg/L}} Q_{\text{ML/d}}}{90 \dfrac{\text{g}}{\text{person}\cdot\text{d}}} \qquad \text{[SI]} \quad 28.3(a)$$

$$P_{e,1000\text{s}} = \frac{\begin{array}{c}\text{BOD}_{\text{mg/L}} Q_{\text{gal/day}} \\ \times \left(8.345 \dfrac{\text{lbm-L}}{\text{MG-mg}}\right)\end{array}}{\left(10^6 \dfrac{\text{gal}}{\text{MG}}\right)(1000 \text{ persons})} \\ \times \left(0.20 \dfrac{\text{lbm}}{\text{person-day}}\right) \qquad \text{[U.S.]} \quad 28.3(b)$$

15. MICROBIAL GROWTH

$$r_g = \frac{dX}{dt} = \mu X \qquad 28.8$$

$$\mu = \mu_m \left(\frac{S}{K_s + S}\right) \qquad 28.9$$

$$r_g = \frac{\mu_m X S}{K_s + S} \qquad 28.10$$

$$r_{\text{su}} = \frac{dS}{dt} = \frac{-\mu_m X S}{Y(K_s + S)} \qquad 28.11$$

$$k = \frac{\mu_m}{Y} \qquad 28.12$$

$$\frac{dS}{dt} = \frac{-kXS}{K_s + S} \qquad 28.13$$

$$r_d = -k_d X \qquad 28.14$$

$$r'_g = \frac{\mu_m X S}{K_s + S} - k_d X$$
$$= -Y r_{\text{su}} - k_d X \qquad 28.15$$

$$\mu' = \mu_m \left(\frac{S}{K_s + S}\right) - k_d \qquad 28.16$$

16. DISSOLVED OXYGEN IN WASTEWATER

$$D = \text{DO}_{\text{sat}} - \text{DO} \qquad 28.17$$

17. REOXYGENATION

$$r_r = K_r(\text{DO}_{\text{sat}} - \text{DO}) \qquad 28.18$$

$$D_t = D_0 10^{-K_r t} \qquad \text{[base-10]} \qquad 28.19$$

$$D_t = D_0 e^{-K'_r t} \qquad \text{[base-}e\text{]} \qquad 28.20$$

$$K'_r = 2.303 K_r \qquad 28.21$$

- O'Connor and Dobbins formula:

$$K'_{r,20°\text{C}} \approx \frac{3.93\sqrt{\text{v}_{\text{m/s}}}}{d_{\text{m}}^{1.5}} \begin{bmatrix} 0.3 \text{ m} < d < 9.14 \text{ m} \\ 0.15 \text{ m/s} < \text{v} < 0.49 \text{ m/s} \end{bmatrix}$$
$$\text{[SI]} \quad 28.22(a)$$

$$K'_{r,68°\text{F}} \approx \frac{12.9\sqrt{\text{v}_{\text{ft/sec}}}}{d_{\text{ft}}^{1.5}} \begin{bmatrix} 1 \text{ ft} < d < 30 \text{ ft} \\ 0.5 \text{ ft/sec} < \text{v} < 1.6 \text{ ft} \end{bmatrix}$$
$$\text{[U.S.]} \quad 28.22(b)$$

- Churchill formula:

$$K'_{r,20°C} \approx \frac{5.049 v_{m/s}^{0.969}}{d_m^{1.67}} \quad \begin{bmatrix} 0.61 \text{ m} < d < 3.35 \text{ m} \\ 0.55 \text{ m/s} < v < 1.52 \text{ m/s} \end{bmatrix}$$

[SI]   *28.23(a)*

$$K'_{r,68°F} \approx \frac{11.61 v_{ft/sec}^{0.969}}{d_{ft}^{1.67}} \quad \begin{bmatrix} 2 \text{ ft} < d < 11 \text{ ft} \\ 1.8 \text{ ft/sec} < v < 5 \text{ ft/sec} \end{bmatrix}$$

[U.S.]   *28.23(b)*

$$K'_{r,T} = K'_{r,20°C}(1.024)^{T-20°C} \qquad 28.24$$

$$K'_{r,T_1} = K'_{r,T_2}\theta_r^{T_1-T_2} \qquad 28.25$$

18. DEOXYGENATION

$$r_{d,t} = -K_d\text{DO} \qquad 28.26$$

$$K'_d = 2.303 K_d \qquad 28.27$$

$$K'_{d,T} = K'_{d,20°C}\theta_d^{T-20°C} \qquad 28.28$$

$$K'_{d,T_1} = K'_{d,T_2}\theta_d^{T_1-T_2} \qquad 28.29$$

20. BIOCHEMICAL OXYGEN DEMAND

$$\text{BOD}_5 = \frac{\text{DO}_i - \text{DO}_f}{\dfrac{V_{\text{sample}}}{V_{\text{sample}} + V_{\text{dilution}}}} \qquad 28.30$$

$$\text{BOD}_t = \text{BOD}_u(1 - 10^{-K_d t})$$
$$= \text{BOD}_u(1 - e^{-K'_d t}) \qquad 28.31$$

$$\text{BOD}_u \approx 1.463 \text{BOD}_5 \qquad 28.32$$

$$\text{BOD}_{T°C} = \text{BOD}_{20°C}(0.02\,T_{°C} + 0.6) \qquad 28.33$$

21. SEEDED BOD

$$\text{BOD} = \frac{\text{DO}_i - \text{DO}_f - x(\text{DO}_i^* - \text{DO}_f^*)}{\dfrac{V_{\text{sample}}}{V_{\text{sample}} + V_{\text{dilution}}}} \qquad 28.34$$

22. DILUTION PURIFICATION

$$C_f = \frac{C_1 Q_1 + C_2 Q_2}{Q_1 + Q_2} \qquad 28.35$$

23. RESPONSE TO DILUTION PURIFICATION

$$D_t = \left(\frac{K_d\text{BOD}_u}{K_r - K_d}\right)(10^{-K_d t} - 10^{-K_r t}) + D_0(10^{-K_r t})$$

28.36

$$x_c = v t_c \qquad 28.37$$

$$t_c = \left(\frac{1}{K_r - K_d}\right)$$
$$\times \log_{10}\left(\left(\frac{K_d\text{BOD}_u - K_r D_0 + K_d D_0}{K_d\text{BOD}_u}\right)\left(\frac{K_r}{K_d}\right)\right)$$

28.38

$$D_c = \left(\frac{K_d\text{BOD}_u}{K_r}\right)10^{-K_d t_c} \qquad 28.39$$

Figure 28.5 *Oxygen Sag Curve*

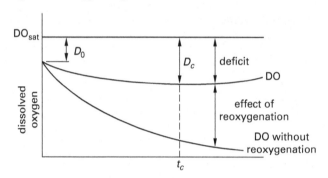

CERM Chapter 29
Wastewater Treatment: Equipment and Processes

11. AERATED LAGOONS

$$t_d = \frac{V}{Q} = \frac{\eta}{k_{1,\text{base-}e}(1 - \eta)} \qquad 29.1$$

$$k_{1,\text{base-}e} = 2.3 k_{1,\text{base-}10} \qquad 29.2$$

13. GRIT CHAMBERS

$$v = \sqrt{8k\left(\frac{g d_p}{f}\right)(\text{SG}_p - 1)} \qquad 29.3$$

17. PLAIN SEDIMENTATION BASINS/CLARIFIERS

$$v^* = \frac{Q}{A} \qquad 29.4$$

$$t_d = \frac{V}{Q} \qquad 29.5$$

$$\text{weir loading} = \frac{Q}{L} \qquad 29.6$$

19. TRICKLING FILTERS

$$S_o + RS_e = (1 + R)S_i \qquad 29.7$$

$$S_i = \frac{S_o + RS_e}{1 + R} \qquad 29.8$$

$$\eta = \frac{S_{\text{removed}}}{S_o} = \frac{S_o - S_e}{S_o} \qquad 29.9$$

$$R = \frac{Q_r}{Q_w} \qquad 29.10$$

$$L_H = \frac{Q_w + Q_r}{A} = \frac{Q_w(1 + R)}{A} \qquad 29.11$$

$$L_{\text{BOD,kg/m}^3\cdot\text{d}} = \frac{Q_{w,\text{m}^3/\text{d}} S_{\text{mg/L}}}{\left(1000 \ \frac{\text{mg}\cdot\text{m}^3}{\text{kg}\cdot\text{L}}\right) V_{\text{m}^3}} \qquad \text{[SI]} \quad 29.12(a)$$

$$L_{\text{BOD,lbm/1000\,ft}^3\text{-day}} = \frac{Q_{w,\text{MGD}} S_{\text{mg/L}} \left(8.345 \ \frac{\text{lbm-L}}{\text{MG-mg}}\right) \times \left(1000 \ \frac{\text{ft}^3}{1000 \ \text{ft}^3}\right)}{V_{\text{ft}^3}}$$
$$\text{[U.S.]} \quad 29.12(b)$$

21. NATIONAL RESEARCH COUNCIL EQUATION

$$\eta = \frac{1}{1 + 0.0561 \sqrt{\dfrac{L_{\text{BOD,lbm/day}}}{V_{\text{1000s ft}^3} F}}} \qquad 29.13$$

$$\eta = \frac{1}{1 + 0.0085 \sqrt{\dfrac{L_{\text{BOD,lbm/day}}}{V_{\text{ac-ft}} F}}} \qquad 29.14$$

$$F = \frac{1 + R}{(1 + wR)^2} \qquad 29.15$$

• *two-stage filter with clarifier*

$$\eta_2 = \frac{1}{1 + \left(\dfrac{0.0561}{1 - \eta_1}\right) \sqrt{\dfrac{L_{\text{BOD}}}{F}}} \qquad 29.16$$

22. VELZ EQUATION

$$\frac{S_e}{S_i} = e^{-KZ} \qquad \left[\text{alternatively } \frac{S_e}{S_i} = 10^{-KZ}\right] \qquad 29.17$$

$$K_T = K_{20°\text{C}}(1.047)^{T-20°\text{C}} \qquad 29.18$$

CERM Chapter 30
Activated Sludge and Sludge Processing

5. SLUDGE PARAMETERS

$$F = S_o Q_o \qquad 30.1$$

$$M = V_a X \qquad 30.2$$

$$\text{F:M} = \frac{S_{o,\text{mg/L}} Q_{o,\text{MGD}}}{V_{a,\text{MG}} X_{\text{mg/L}}} = \frac{S_{o,\text{mg/L}}}{\theta_{\text{days}} X_{\text{mg/L}}} \qquad 30.3$$

$$\text{F:M} = \frac{S_{o,\text{mg/L}} Q_{o,\text{MGD}}}{V_{a,\text{MG}} \text{MLSS}_{\text{mg/L}}} \qquad 30.4$$

$$\theta_c = \frac{V_a X}{Q_e X_e + Q_w X_w} \qquad 30.5$$

$$\theta_{\text{BOD}} = \frac{1}{\text{F:M}} = \frac{V_{a,\text{m}^3} X_{\text{mg/L}}}{S_{o,\text{mg/L}} Q_{o,\text{m}^3/\text{d}}} \qquad \text{[SI]} \quad 30.6(a)$$

$$\theta_{\text{BOD}} = \frac{1}{\text{F:M}} = \frac{V_{a,\text{MG}} X_{\text{mg/L}}}{S_{o,\text{mg/L}} Q_{o,\text{MGD}}} \qquad \text{[U.S.]} \quad 30.6(b)$$

$$\text{SVI}_{\text{mg/L}} = \frac{\left(1000 \ \dfrac{\text{mg}}{\text{g}}\right) V_{\text{settled,mL/L}}}{\text{MLSS}_{\text{mg/L}}} \qquad 30.7$$

$$\text{SSV} = \frac{V_{\text{settled sludge,mL}} \left(1000 \ \dfrac{\text{mL}}{\text{L}}\right)}{V_{\text{sample,mL}}} \qquad 30.8$$

$$\text{TSS}_{\text{mg/L}} = \frac{\left(1000 \ \dfrac{\text{mg}}{\text{g}}\right)\left(1000 \ \dfrac{\text{mL}}{\text{L}}\right)}{\text{SVI}_{\text{mL/g}}} \qquad 30.9$$

6. SOLUBLE BOD ESCAPING TREATMENT

$$\text{BOD}_e = \text{BOD}_{\text{escaping treatment}} + \text{BOD}_{\text{effluent suspended solids}}$$
$$= S + S_e$$
$$= S + 1.42fGX_e \qquad 30.10$$

$$f = \frac{\text{BOD}_5}{\text{BOD}_u} \qquad 30.11$$

7. PROCESS EFFICIENCY

$$\eta_{\text{BOD}} = \frac{S_o - S}{S_o} \qquad\qquad 30.12$$

8. PLUG FLOW AND STIRRED TANK MODELS

$$\frac{1}{\theta_c} = \frac{\mu_m(S_o - S)}{S_o - S + (1 + R)K_s \ln\frac{S_i}{S}} - k_d \quad \text{[PFR only]}$$

$$\qquad\qquad 30.13$$

$$\mu_m = kY \qquad\qquad 30.14$$

$$S_i = \frac{S_o + RS}{1 + R} \quad \text{[PFR only]} \qquad 30.15$$

$$r_{\text{su}} = \frac{-\mu_m SX}{Y(K_s + S)} \quad \text{[PFR only]} \qquad 30.16$$

$$U = \eta(\text{F:M}) = \frac{-r_{\text{su}}}{X}$$

$$= \frac{S_o - S}{\theta X} \qquad\qquad 30.17$$

$$\frac{1}{\theta_c} = Y(\text{F:M})\eta - k_d$$

$$= -Y\left(\frac{r_{\text{su}}}{X}\right) - k_d$$

$$= YU - k_d \quad \text{[CSTR only]} \qquad 30.18$$

$$S = \frac{K_s(1 + k_d\theta_c)}{\theta_c(\mu_m - k_d) - 1} \quad \text{[CSTR only]} \qquad 30.19$$

$$\theta = \frac{V_a}{Q_o} \quad \text{[PFR and CSTR]} \qquad 30.20$$

$$\theta_s = \frac{V_a + V_s}{Q_o} \quad \text{[PFR and CSTR]} \qquad 30.21$$

$$V_a = \theta Q_o$$

$$= \frac{\theta_c Q_o Y(S_o - S)}{X(1 + k_d\theta_c)} \quad \text{[PFR and CSTR]} \qquad 30.22$$

$$X = \frac{\left(\dfrac{\theta_c}{\theta}\right) Y(S_o - S)}{1 + k_d\theta_c} \quad \text{[PFR and CSTR]} \qquad 30.23$$

$$Y_{\text{obs}} = \frac{Y}{1 + k_d\theta_c} \quad \text{[PFR and CSTR]} \qquad 30.24$$

$$P_{x,\text{kg/d}} = \frac{Q_{w,\text{m}^3/\text{d}} X_{r,\text{mg/L}}}{1000 \, \dfrac{\text{g}}{\text{kg}}}$$

$$= \frac{Y_{\text{obs,mg/mg}} Q_{o,\text{m}^3/\text{d}}(S_o - S)_{\text{mg/L}}}{1000 \, \dfrac{\text{g}}{\text{kg}}}$$

$$\text{[SI; PFR and CSTR]} \qquad 30.25(a)$$

$$P_{x,\text{lbm/day}} = Q_{w,\text{MGD}} X_{r,\text{mg/L}}\left(8.345 \, \frac{\text{lbm-L}}{\text{MG-mg}}\right)$$

$$= Y_{\text{obs,lbm/lbm}} Q_{o,\text{MGD}}(S_o - S)_{\text{mg/L}}$$

$$\times \left(8.345 \, \frac{\text{lbm-L}}{\text{MG-mg}}\right)$$

$$\text{[U.S.; PFR and CSTR]} \qquad 30.25(b)$$

$$m_{e,\text{solids}} = Q_e X_e \quad \text{[PFR and CSTR]} \qquad 30.26$$

$$\theta_c = \frac{V_a X}{Q_w X_r + Q_e X_e}$$

$$= \frac{V_a X}{Q_w X_r + (Q_o - Q_w)X_e} \quad \text{[PFR and CSTR]} \qquad 30.27$$

$$\theta_c \approx \frac{V_a X}{Q_w X_r} \quad \text{[PFR and CSTR]} \qquad 30.28$$

$$\theta_c \approx \frac{V_a}{Q_w} \quad \text{[CSTR]} \qquad 30.29$$

10. AERATION TANKS

$$\theta = \frac{V_a}{Q_o} \qquad\qquad 30.30$$

$$L_{\text{BOD}} = \frac{S_o Q_o}{V_a\left(1000 \, \dfrac{\text{mg·m}^3}{\text{kg·L}}\right)} \qquad \text{[SI]} \quad 30.31(a)$$

$$L_{\text{BOD}} = \frac{S_{o,\text{mg/L}} Q_{o,\text{MGD}}\left(8.345 \, \dfrac{\text{lbm-L}}{\text{MG-mg}}\right)(1000)}{V_{a,\text{ft}^3}}$$

$$\text{[U.S.]} \quad 30.31(b)$$

$$\dot{m}_{\text{oxygen}} = K_t D \qquad\qquad 30.32$$

$$D = \beta \text{DO}_{\text{saturated water}} - \text{DO}_{\text{mixed liquor}} \qquad 30.33$$

$$\dot{m}_{\text{oxygen,kg/d}} = \frac{Q_{o,\text{m}^3/\text{d}}(S_o - S)_{\text{mg/L}}}{f\left(1000 \ \frac{\text{mg·m}^3}{\text{kg·L}}\right)} - 1.42 P_{x,\text{kg/d}}$$

$$[\text{SI}] \quad 30.34(a)$$

$$\dot{m}_{\text{oxygen,lbm/day}} = \frac{Q_{o,\text{MGD}}(S_o - S)_{\text{mg/L}}\left(8.345 \ \frac{\text{lbm-L}}{\text{MG-mg}}\right)}{f}$$
$$- 1.42 P_{x,\text{lbm/day}}$$

$$[\text{U.S.}] \quad 30.34(b)$$

$$\dot{m}_{\text{air}} = \frac{\dot{m}_{\text{oxygen}}}{0.232\eta_{\text{transfer}}} \qquad 30.35$$

$$\dot{V}_{\text{air}} = \frac{\dot{m}_{\text{air}}}{\rho_{\text{air}}} \quad [\text{PFR and CSTR}] \qquad 30.36$$

$$V_{\text{air,m}^3/\text{kg BOD}} = \frac{\dot{V}_{\text{air,m}^3/\text{d}}\left(1000 \ \frac{\text{mg·m}^3}{\text{kg·L}}\right)}{Q_{o,\text{m}^3/\text{d}}(S_o - S)_{\text{mg/L}}} \quad [\text{SI}] \quad 30.37(a)$$

$$V_{\text{air,ft}^3/\text{lbm BOD}} = \frac{\dot{V}_{\text{air,ft}^3/\text{day}}}{Q_{o,\text{MGD}}(S_o - S)_{\text{mg/L}}\left(8.345 \ \frac{\text{lbm-L}}{\text{MG-mg}}\right)}$$

$$[\text{U.S.}] \quad 30.37(b)$$

11. AERATION POWER AND COST

$$P_{\text{ideal,kW}} = -\left(\frac{kp_1\dot{V}_1}{(k-1)\left(1000 \ \frac{\text{W}}{\text{kW}}\right)}\right)\left(1 - \left(\frac{p_2}{p_1}\right)^{(k-1)/k}\right)$$

$$= -\left(\frac{k\dot{m}R_{\text{air}}T_1}{(k-1)\left(1000 \ \frac{\text{W}}{\text{kW}}\right)}\right)\left(1 - \left(\frac{p_2}{p_1}\right)^{(k-1)/k}\right)$$

$$= -\left(\frac{c_p\dot{m}T_1}{1000 \ \frac{\text{W}}{\text{kW}}}\right)\left(1 - \left(\frac{p_2}{p_1}\right)^{(k-1)/1}\right)$$

$$[\text{SI}] \quad 30.38(a)$$

$$P_{\text{ideal,hp}} = -\left(\frac{kp_1\dot{V}_1}{(k-1)\left(550 \ \frac{\text{ft-lbf}}{\text{hp-sec}}\right)}\right)\left(1 - \left(\frac{p_2}{p_1}\right)^{(k-1)/k}\right)$$

$$= -\left(\frac{k\dot{m}R_{\text{air}}T_1}{(k-1)\left(550 \ \frac{\text{ft-lbf}}{\text{hp-sec}}\right)}\right)\left(1 - \left(\frac{p_2}{p_1}\right)^{(k-1)/k}\right)$$

$$= -\left(\frac{c_p J\dot{m}T_1}{550 \ \frac{\text{ft-lbf}}{\text{hp-sec}}}\right)\left(1 - \left(\frac{p_2}{p_1}\right)^{(k-1)/k}\right)$$

$$[\text{U.S.}] \quad 30.38(b)$$

$$P_{\text{actual}} = \frac{P_{\text{ideal}}}{\eta_c} \qquad 30.39$$

$$\text{total cost} = C_{\text{kW-hr}}P_{\text{actual}}t \qquad 30.40$$

12. RECYCLE RATIO AND RECIRCULATION RATE

$$R = \frac{Q_r}{Q_o} \qquad 30.41$$

$$X_o Q_o + X_r Q_r = X(Q_o + Q_r) \qquad 30.42$$

$$\frac{Q_r}{Q_o + Q_r} = \frac{X}{X_r} \qquad 30.43$$

$$\frac{Q_r}{Q_o + Q_r} = \frac{V_{\text{settled,mL/L}}}{1000 \ \frac{\text{mL}}{\text{L}}} \qquad 30.44$$

$$R = \frac{V_{\text{settled,mL/L}}}{1000 \ \frac{\text{mL}}{\text{L}} - V_{\text{settled,mL/L}}}$$

$$= \frac{1}{\dfrac{10^6}{(\text{SVI}_{\text{mL/g}})(\text{MLSS}_{\text{mg/L}})} - 1} \qquad 30.45$$

15. QUANTITIES OF SLUDGE

$$m_{\text{wet}} = V\rho_{\text{sludge}} = V(\text{SG}_{\text{sludge}})\rho_{\text{water}} \qquad 30.46$$

$$\frac{1}{\text{SG}_{\text{sludge}}} = \frac{1-s}{1} + \frac{s}{\text{SG}_{\text{solids}}}$$

$$= \frac{1 - s_{\text{fixed}} - s_{\text{volatile}}}{1} + \frac{s_{\text{fixed}}}{\text{SG}_{\text{fixed solids}}}$$

$$+ \frac{s_{\text{volatile}}}{\text{SG}_{\text{volatile solids}}} \qquad 30.47$$

$$V_{\text{sludge,wet}} = \frac{m_{\text{dried}}}{s\rho_{\text{sludge}}} \approx \frac{m_{\text{dried}}}{s\rho_{\text{water}}} \qquad 30.48$$

$$m_{\text{dried,kg/d}} = \frac{(\Delta\text{SS})_{\text{mg/L}} Q_{o,\text{m}^3/\text{d}}}{1000 \ \dfrac{\text{mg}\cdot\text{m}^3}{\text{kg}\cdot\text{L}}} \qquad [\text{SI}] \quad 30.49(a)$$

$$m_{\text{dried,lbm/day}} = (\Delta\text{SS})_{\text{mg/L}} Q_{o,\text{MGD}} \left(8.345 \ \frac{\text{lbm-L}}{\text{MG-mg}}\right)$$

$$[\text{U.S.}] \quad 30.49(b)$$

$$m_{\text{dried,kg/d}} = \frac{K S_{o,\text{mg/L}} Q_{o,\text{m}^3/\text{d}}}{1000 \ \dfrac{\text{mg}\cdot\text{m}^3}{\text{kg}\cdot\text{L}}}$$

$$= \frac{Y_{\text{obs}}(S_o - S)_{\text{mg/L}} Q_{o,\text{m}^3/\text{d}}}{1000 \ \dfrac{\text{mg}\cdot\text{m}^3}{\text{kg}\cdot\text{L}}} \qquad [\text{SI}] \quad 30.50(a)$$

$$m_{\text{dried,lbm/day}} = K S_{o,\text{mg/L}} Q_{o,\text{MGD}} \left(8.345 \ \frac{\text{lbm-L}}{\text{MG-mg}}\right)$$

$$= Y_{\text{obs}}(S_o - S)_{\text{mg/L}} Q_{o,\text{MGD}}$$

$$\times \left(8.345 \ \frac{\text{lbm-L}}{\text{MG-mg}}\right)$$

$$[\text{U.S.}] \quad 30.50(b)$$

20. METHANE PRODUCTION

$$V_{\text{methane,m}^3/\text{d}} = \left(0.35 \ \frac{\text{m}^3}{\text{kg}}\right) \left(\frac{E S_{o,\text{mg/L}} Q_{\text{m}^3/\text{d}}}{1000 \ \dfrac{\text{g}}{\text{kg}}} - 1.42 P_{x,\text{kg/d}} \right)$$

$$[\text{SI}] \quad 30.51(a)$$

$$V_{\text{methane,ft}^3/\text{day}} = \left(5.61 \ \frac{\text{ft}^3}{\text{lbm}}\right) \left(\begin{array}{c} E S_{o,\text{mg/L}} Q_{\text{MGD}} \\ \times \left(8.345 \ \dfrac{\text{lbm-L}}{\text{MG-mg}}\right) \\ - 1.42 P_{x,\text{lbm/day}} \end{array} \right)$$

$$[\text{U.S.}] \quad 30.51(b)$$

$$q = (\text{LHV}) V_{\text{fuel}} \qquad 30.52$$

21. HEAT TRANSFER AND LOSS

- *sensible energy*

$$q = m_{\text{sludge}} c_p (T_2 - T_1)$$

$$= V_{\text{sludge}} \rho_{\text{sludge}} c_p (T_2 - T_1) \qquad 30.53$$

- *convective energy*

$$q = h A_{\text{surface}} (T_{\text{surface}} - T_{\text{air}}) \qquad 30.54$$

- *total energy*

$$q = U A_{\text{surface}} (T_{\text{contents}} - T_{\text{air}}) \qquad 30.55$$

- *conductive energy*

$$q = \frac{k A (T_{\text{inner}} - T_{\text{outer}})}{L} \qquad 30.56$$

22. SLUDGE DEWATERING

$$V_{\text{press}} \rho_{\text{filter cake}} s_{\text{filter cake}}$$

$$= V_{\text{sludge,per cycle}} s_{\text{sludge}} \rho_{\text{sludge}}$$

$$= V_{\text{sludge,per cycle}} s_{\text{sludge}} \rho_{\text{water}} \text{SG}_{\text{sludge}} \qquad 30.57$$

- *centrifuge*

$$G = \frac{\omega^2 r}{g} = \left(\frac{2\pi n}{60}\right)^2 \left(\frac{r}{g}\right) \qquad 30.58$$

CERM Chapter 31
Municipal Solid Waste

3. LANDFILL CAPACITY

$$V_c = (\text{CF}) V_o \qquad 31.1$$

$$\Delta V = \frac{NG(\text{LF})}{\gamma} = \frac{NG(\text{LF})}{\rho g} \qquad [\text{SI}] \quad 31.2(a)$$

$$\Delta V = \frac{NG(\text{LF})}{\gamma} = \frac{NG(\text{LF}) g_c}{\rho g} \qquad [\text{U.S.}] \quad 31.2(b)$$

$$\text{LF} = \frac{V_{\text{MSW}} + V_{\text{cover soil}}}{V_{\text{MSW}}} \qquad 31.3$$

11. LANDFILL GAS

$$p_t = p_{\text{CH}_4} + p_{\text{CO}_2} + p_{\text{H}_2\text{O}} + p_{\text{N}_2} + p_{\text{other}} \qquad 31.8$$

$$p_i = B_i p_t \qquad 31.9$$

13. LEACHATE MIGRATION FROM LANDFILLS

$$Q = KiA \qquad 31.13$$

$$i = \frac{dH}{dL} \qquad 31.14$$

19. INCINERATION OF MUNICIPAL SOLID WASTE

$$\text{HRR} = \frac{(\text{fueling rate})(\text{HV})}{\text{total effective grate area}} \qquad 31.24$$

CERM Chapter 32
Pollutants in the Environment

37. SMOKE

$$\text{optical density} = \log_{10} \frac{1}{1 - \text{opacity}} \qquad 32.7$$

CERM Chapter 34
Environmental Remediation

12. BIOFILTRATION

$$\eta = 1 - \frac{C_{\text{out}}}{C_{\text{in}}} = 1 - e^{-kt} \qquad 34.14$$

16. CYCLONE SEPARATORS

$$S = \frac{\text{v}_{\text{inlet}}^2}{rg} \qquad 34.15$$

$$h = \frac{KBH\text{v}_{\text{inlet}}^2}{2gD_e^2} \qquad 34.16$$

18. ELECTROSTATIC PRECIPITATORS

$$\eta = \frac{C_{\text{in}} - C_{\text{out}}}{C_{\text{in}}} = \frac{\dot{m}_{\text{in}} - \dot{m}_{\text{out}}}{\dot{m}_{\text{in}}} = 1 - \exp\left(\frac{-Aw_e}{Q}\right)^y \qquad 34.22$$

47. STRIPPING, AIR

$$p_{\text{A}} = H_{\text{A}} x_{\text{A}} \qquad 34.43$$

$$R = \frac{HG}{L} \qquad 34.44$$

Geotechnical

CERM Chapter 35
Soil Properties and Testing

1. SOIL PARTICLE SIZE DISTRIBUTION

$$C_u = \frac{D_{60}}{D_{10}} \qquad 35.1$$

$$C_z = \frac{D_{30}^2}{D_{10}D_{60}} \qquad 35.2$$

3. AASHTO SOIL CLASSIFICATION

$$I_g = (F_{200} - 35)\big(0.2 + 0.005(\text{LL} - 40)\big)$$
$$+ 0.01(F_{200} - 15)(\text{PI} - 10) \qquad 35.3$$

5. MASS-VOLUME RELATIONSHIPS

(See Table 35.7.)

$$D_r = \frac{e_{\max} - e}{e_{\max} - e_{\min}} \qquad 35.16$$

7. EFFECTIVE STRESS

$$\sigma' = \sigma - u \qquad 35.17$$

10. CONE PENETROMETER TEST

$$f_R = \frac{q_s}{q_c} \times 100\% \qquad 35.18$$

11. PROCTOR TEST

$$\text{RC} = \frac{\rho_d}{\rho_d^*} \times 100\% \qquad 35.19$$

12. MODIFIED PROCTOR TEST

$$\rho_z = \frac{m_s}{V_w + V_s} \qquad 35.20$$

$$\rho_z = \frac{\rho_w}{w + \dfrac{1}{\text{SG}}} \qquad 35.21$$

$$\rho_s = (\text{SG})\rho_w \qquad 35.22$$

14. ATTERBERG LIMIT TESTS

$$\text{PI} = \text{LL} - \text{PL} \qquad 35.23$$
$$\text{SI} = \text{PL} - \text{SL} \qquad 35.24$$
$$\text{LI} = \frac{w - \text{PL}}{\text{PI}} \qquad 35.25$$

15. PERMEABILITY TESTS

(See Fig. 35.8.)

$$Q = \text{v}A_{\text{gross}} \qquad 35.26$$
$$\text{v} = Ki \qquad 35.27$$
$$K_{\text{cm/s}} \approx CD_{10,\text{mm}}^2 \qquad 35.28$$

$$K = \frac{VL}{hAt} \quad \text{[constant head]} \qquad 35.29$$

$$K = \frac{A'L}{At} \ln \frac{h_i}{h_f} \quad \text{[falling head]} \qquad 35.30$$

- *auger hole*

$$F = \frac{11r}{2} \qquad 35.31$$

$$K = \frac{\pi r^2}{Ft} \ln \frac{h_i}{h_f}$$
$$= \frac{2\pi r}{11t} \ln \frac{h_i}{h_f} \quad \text{[auger hole]} \qquad 35.32$$

Table 35.7 Soil Indexing Formulas

property	saturated sample $(m_s, m_w, SG$ are known$)$	unsaturated sample $(m_s, m_w, SG, V_t$ are known$)$	supplementary formulas relating measured and computed factors				
volume components							
V_s volume of solids		$\frac{m_s}{(SG)\rho_w}$	$V_t - (V_g + V_w)$	$V_t(1-n)$	$\frac{V_t}{1+e}$	$\frac{V_v}{e}$	
V_w volume of water		$\frac{m_w}{\rho_w^*}$	$V_v - V_g$	SV_v	$\frac{SV_te}{1+e}$	SV_se	
V_g volume of gas or air	zero	$V_t - (V_s + V_w)$	$V_v - V_w$	$(1-S)V_v$	$\frac{(1-S)V_te}{1+e}$	$(1-S)V_se$	
V_v volume of voids	$\frac{m_w}{\rho_w^*}$	$V_t - \frac{m_s}{(SG)\rho_w}$	$V_t - V_s$	$\frac{V_sn}{1-n}$	$\frac{V_te}{1+e}$	V_se	
V_t total volume of sample	$V_s + V_w$	measured $(V_g + V_w + V_s)$	$V_s + V_g + V_w$	$\frac{V_s}{1-n}$	$V_s(1+e)$	$\frac{V_v(1+e)}{e}$	
n porosity		$\frac{V_v}{V_t}$	$1 - \frac{V_s}{V_t}$	$1 - \frac{m_s}{(SG)V_t\rho_w}$	$\frac{e}{1+e}$		
e void ratio		$\frac{V_v}{V_s}$	$\frac{V_t}{V_s} - 1$	$\frac{(SG)V_t\rho_w}{m_s} - 1$	$\frac{m_w(SG)}{m_sS}$	$\frac{n}{1-n}$	$w\left(\frac{SG}{S}\right)$
mass for specific sample							
m_s mass of solids		measured	$\frac{m_t}{1+w}$	$(SG)V_t\rho_w(1-n)$	$\frac{m_w(SG)}{eS}$	$V_s(SG)\rho_w$	
m_w mass of water		measured	wm_s	$\frac{m_tw}{1+w}$	$S\rho_wVv$	$\frac{em_sS}{SG}$	$V_t\rho_dw$
m_t total mass of sample		$m_s + m_w$	$m_s(1+w)$				
mass for sample of unit volume (density)							
ρ_d dry density	$\frac{m_s}{V_s + V_w}$	$\frac{m_s}{V_g + V_w + V_s}$	$\frac{m_t}{V_t(1+w)}$	$\frac{(SG)\rho_w}{1+e}$	$\frac{(SG)\rho_w}{1+\frac{w(SG)}{S}}$	$\frac{\rho}{1+w}$	
ρ wet density (density with moisture)	$\frac{m_s + m_w}{V_s + V_w}$	$\frac{m_s + m_w}{V_t}$	$\frac{m_t}{V_t}$	$\frac{(SG + Se)\rho_w}{1+e}$	$\frac{(1+w)\rho_w}{\frac{w}{S}+\frac{1}{SG}}$	$\rho_d(1+w)$	
ρ_{sat} saturated density	$\frac{m_s + m_w}{V_s + V_w}$	$\frac{m_s + V_v\rho_w}{V_t}$	$\frac{m_s}{V_t} + \left(\frac{e}{1+e}\right)\rho_w$	$\frac{(SG + e)\rho_w}{1+e}$	$\frac{(1+w)\rho_w}{w+\frac{1}{SG}}$		
ρ_b buoyant (submerged) density		$\rho_{sat} - \rho_w^*$	$\frac{m_s}{V_t} - \left(\frac{1}{1+e}\right)\rho_w^*$	$\left(\frac{SG+e}{1+e} - 1\right)\rho_w^*$	$\left(\frac{1-\frac{1}{SG}}{w+\frac{1}{SG}}\right)\rho_w^*$		
combined relations							
w water content		$\frac{m_w}{m_s}$	$\frac{m_t}{m_s} - 1$	$\frac{Se}{SG}$	$S\left(\frac{\rho_w^*}{\rho_d} - \frac{1}{SG}\right)$		
S degree of saturation	100%	$\frac{V_w}{V_v}$	$\frac{m_w}{V_v\rho_w}$	$\frac{w(SG)}{e}$	$\frac{w}{\frac{\rho_w^*}{\rho_d} - \frac{1}{SG}}$		
SG specific gravity of solids		$\frac{m_s}{V_s\rho_w}$	$\frac{Se}{w}$				

ρ_w is the density of water. Where noted with an asterisk (*), use the actual density of water at the recorded temperature. In other cases, use 62.4 lbm/ft³ or 1000 kg/m³.

Figure 35.8 *Permeameters*

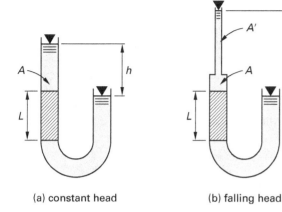

(a) constant head   (b) falling head

16. CONSOLIDATION TESTS

$$\text{OCR} = \frac{p'_{\max}}{p'_o} \qquad 35.33$$

$$C_c = -\frac{e_1 - e_2}{\log_{10}\dfrac{p'_1}{p'_2}} \qquad 35.34$$

$$C_{\epsilon c} = \frac{C_c}{1 + e_0} \qquad 35.35$$

$$C_c \approx 0.009(\text{LL} - 10) \qquad 35.36$$

Figure 35.10 *e-log p Curve*

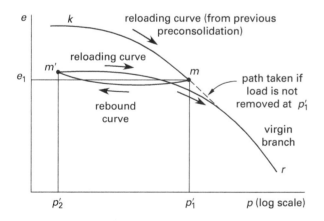

17. DIRECT SHEAR TEST

$$S = \tau = c + \sigma \tan \phi \qquad 35.37$$

18. TRIAXIAL STRESS TEST

$$\sigma_\alpha = \tfrac{1}{2}(\sigma_A + \sigma_R) + \tfrac{1}{2}(\sigma_A - \sigma_R)\cos 2\alpha \qquad 35.38$$

$$\tau_\alpha = \tfrac{1}{2}(\sigma_A - \sigma_R)\sin 2\alpha \qquad 35.39$$

$$\frac{\sigma_1}{\sigma_3} = \frac{1 + \sin \phi}{1 - \sin \phi} \quad [c = 0] \qquad 35.40$$

$$\alpha = 45^\circ + \tfrac{1}{2}\phi \qquad 35.41$$

$$S_{\text{us}} = c = \frac{\sigma_D}{2} \qquad 35.42$$

$$s = c' + \sigma' \tan \phi' \qquad 35.43$$

20. UNCONFINED COMPRESSIVE STRENGTH TEST

$$S_{\text{uc}} = \frac{P}{A} \qquad 35.44$$

$$s_u = \frac{S_{\text{uc}}}{2} \qquad 35.45$$

21. SENSITIVITY

$$S_t = \frac{S_{\text{undisturbed}}}{S_{\text{remolded}}} \qquad 35.46$$

22. CALIFORNIA BEARING RATIO TEST

$$\text{CBR} = \frac{\text{actual load}}{\text{standard load}} \times 100 \qquad 35.47$$

26. CHARACTERIZING ROCK MASS QUALITY

$$\text{RQD} = \frac{\displaystyle\sum_{L_i > 4\text{ in}} L_i}{L_{\text{core}}} \times 100\% \qquad 35.48$$

CERM Chapter 36
Shallow Foundations

5. GENERAL BEARING CAPACITY EQUATION

$$q_{\text{ult}} = \tfrac{1}{2}\rho g B N_\gamma + c N_c + \left(p_q + \rho g D_f\right) N_q \quad \text{[SI]} \qquad 36.1(a)$$

$$q_{\text{ult}} = \tfrac{1}{2}\gamma B N_\gamma + c N_c + \left(p_q + \gamma D_f\right) N_q \quad \text{[U.S.]} \qquad 36.1(b)$$

$$d_c = 1 + \frac{K D_f}{B} \qquad 36.2$$

$$q_{net} = q_{ult} - \rho g D_f \qquad \text{[SI]} \quad 36.3(a)$$

$$q_{net} = q_{ult} - \gamma D_f \qquad \text{[U.S.]} \quad 36.3(b)$$

$$q_a = \frac{q_{net}}{F} \qquad 36.4$$

Table 36.2 *Terzaghi Bearing Capacity Factors for General Shear**

ϕ	N_c	N_q	N_γ
0°	5.7	1.0	0.0
5°	7.3	1.6	0.5
10°	9.6	2.7	1.2
15°	12.9	4.4	2.5
20°	17.7	7.4	5.0
25°	25.1	12.7	9.7
30°	37.2	22.5	19.7
34°	52.6	36.5	35.0
35°	57.8	41.4	42.4
40°	95.7	81.3	100.4
45°	172.3	173.3	297.5
48°	258.3	287.9	780.1
50°	347.5	415.1	1153.2

*Curvilinear interpolation may be used. Do not use linear interpolation.

Table 36.3 *Meyerhof and Vesic Bearing Capacity Factors for General Shear[a]*

ϕ	N_c	N_q	N_γ	N_γ[b]
0°	5.14	1.0	0.0	0.0
5°	6.5	1.6	0.07	0.5
10°	8.3	2.5	0.37	1.2
15°	11.0	3.9	1.1	2.6
20°	14.8	6.4	2.9	5.4
25°	20.7	10.7	6.8	10.8
30°	30.1	18.4	15.7	22.4
32°	35.5	23.2	22.0	30.2
34°	42.2	29.4	31.2	41.1
36°	50.6	37.7	44.4	56.3
38°	61.4	48.9	64.1	78.0
40°	75.3	64.2	93.7	109.4
42°	93.7	85.4	139.3	155.6
44°	118.4	115.3	211.4	224.6
46°	152.1	158.5	328.7	330.4
48°	199.3	222.3	526.5	496.0
50°	266.9	319.1	873.9	762.9

[a]Curvilinear interpolation may be used. Do not use linear interpolation.
[b]This is predicted by the Vesic equation, $N_\gamma = 2(N_q + 1)\tan\phi$.

Table 36.4 *N_c Bearing Capacity Factor Multipliers for Various Values of B/L*

B/L	multiplier
1 (square)	1.25
0.5	1.12
0.2	1.05
0.0	1.00
1 (circular)	1.20

Table 36.5 *N_γ Multipliers for Various Values of B/L*

B/L	multiplier
1 (square)	0.85
0.5	0.90
0.2	0.95
0.0	1.00
1 (circular)	0.70

7. BEARING CAPACITY OF CLAY

$$S_u = c = \frac{S_{uc}}{2} \qquad 36.5$$

$$q_{ult} = cN_c + \rho g D_f \qquad \text{[SI]} \quad 36.6(a)$$

$$q_{ult} = cN_c + \gamma D_f \qquad \text{[U.S.]} \quad 36.6(b)$$

$$q_{net} = q_{ult} - \rho g D_f = cN_c \qquad \text{[SI]} \quad 36.7(a)$$

$$q_{net} = q_{ult} - \gamma D_f = cN_c \qquad \text{[U.S.]} \quad 36.7(b)$$

$$q_a = \frac{q_{net}}{F} \qquad 36.8$$

8. BEARING CAPACITY OF SAND

$$q_{ult} = \tfrac{1}{2}B\rho g N_\gamma + (p_q + \rho g D_f)N_q \qquad \text{[SI]} \quad 36.9(a)$$

$$q_{ult} = \tfrac{1}{2}B\gamma N_\gamma + (p_q + \gamma D_f)N_q \qquad \text{[U.S.]} \quad 36.9(b)$$

$$\begin{aligned} q_{net} &= q_{ult} - \rho g D_f \\ &= \tfrac{1}{2}B\rho g N_\gamma + \rho g D_f(N_q - 1) \end{aligned} \qquad \text{[SI]} \quad 36.10(a)$$

$$\begin{aligned} q_{net} &= q_{ult} - \gamma D_f \\ &= \tfrac{1}{2}B\gamma N_\gamma + \gamma D_f(N_q - 1) \end{aligned} \qquad \text{[U.S.]} \quad 36.10(b)$$

$$\begin{aligned} q_a &= \frac{q_{net}}{F} \\ &= \frac{B}{F}\left(\tfrac{1}{2}\rho g N_\gamma + \rho g(N_q - 1)\left(\frac{D_f}{B}\right)\right) \end{aligned} \qquad \text{[SI]} \quad 36.11(a)$$

$$\begin{aligned} q_a &= \frac{q_{net}}{F} \\ &= \frac{B}{F}\left(\tfrac{1}{2}\gamma N_\gamma + \gamma(N_q - 1)\left(\frac{D_f}{B}\right)\right) \end{aligned} \qquad \text{[U.S.]} \quad 36.11(b)$$

$$q_a = 0.11 C_n N \quad [\text{in tons/ft}^2, B > 2\text{-}4 \text{ ft}, N \le 50] \qquad 36.12$$

9. SHALLOW WATER TABLE CORRECTION

$$C_w = 0.5 + 0.5\left(\frac{D_w}{D_f + B}\right) \qquad 36.13$$

11. EFFECTS OF WATER TABLE ON FOOTING DESIGN

- sand; water table at base of footing; $c = 0$:

$$q_{\text{ult}} = \tfrac{1}{2}\rho_b g B N_\gamma + \rho g D_f N_q \qquad \text{[SI]} \qquad 36.14(a)$$

$$q_{\text{ult}} = \tfrac{1}{2}\gamma_b B N_\gamma + \gamma D_f N_q \qquad \text{[U.S.]} \qquad 36.14(b)$$

- sand; water table at surface; $c = 0$:

$$q_{\text{ult}} = \tfrac{1}{2}\rho_b g B N_\gamma + \rho_b g D_f N_q \qquad \text{[SI]} \qquad 36.15(a)$$

$$q_{\text{ult}} = \tfrac{1}{2}\gamma_b B N_\gamma + \gamma_b D_f N_q \qquad \text{[U.S.]} \qquad 36.15(b)$$

- sand; water table between base and surface; $c = 0$:

$$\left(\rho g D_w + \left(\rho - 1000\ \frac{\text{kg}}{\text{m}^3}\right) g (D_f - D_w)\right) N_q$$
$$= \left(\rho g D_f + \left(1000\ \frac{\text{kg}}{\text{m}^3}\right) g (D_w - D_f)\right) N_q$$
$$\text{[SI]} \qquad 36.16(a)$$

$$\left(\gamma D_w + \left(\gamma - 62.4\ \frac{\text{lbf}}{\text{ft}^3}\right)(D_f - D_w)\right) N_q$$
$$= \left(\gamma D_f + \left(62.4\ \frac{\text{lbf}}{\text{ft}^3}\right)(D_w - D_f)\right) N_q$$
$$\text{[U.S.]} \qquad 36.16(b)$$

- sand; water table below base of footing; $c = 0$:

$$q_{\text{ult}} = \tfrac{1}{2}\rho g B N_\gamma + \rho g D_f N_q \qquad \text{[SI]} \qquad 36.17(a)$$

$$q_{\text{ult}} = \tfrac{1}{2}\gamma B N_\gamma + \gamma D_f N_q \qquad \text{[U.S.]} \qquad 36.17(b)$$

12. ECCENTRIC LOADS ON RECTANGULAR FOOTINGS

$$\epsilon_B = \frac{M_B}{P}; \ \epsilon_L = \frac{M_L}{P} \qquad 36.18$$

$$L' = L - 2\epsilon_L; \ B' = B - 2\epsilon_B \qquad 36.19$$

$$A' = L'B' \qquad 36.20$$

$$p_{\max}, p_{\min} = \frac{P}{BL}\left(1 \pm \frac{6\epsilon}{B}\right) \qquad 36.21$$

Figure 36.6 *Footing with Overturning Moment*

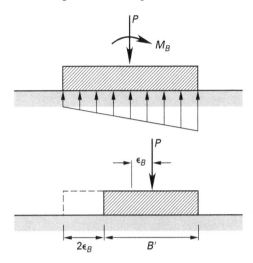

14. RAFTS ON CLAY

$$F = \frac{cN_c}{\dfrac{P_{\text{total}}}{A_{\text{raft}}} - \rho g D_f} \qquad \text{[SI]} \qquad 36.22(a)$$

$$F = \frac{cN_c}{\dfrac{P_{\text{total}}}{A_{\text{raft}}} - \gamma D_f} \qquad \text{[U.S.]} \qquad 36.22(b)$$

15. RAFTS ON SAND

$$q_a = 0.22 C_w C_n N \quad [\text{in tons/ft}^2] \qquad 36.23$$

$$p = \frac{P_{\text{total}}}{A_{\text{raft}}} - \rho g D_f \qquad \text{[SI]} \qquad 36.24(a)$$

$$p = \frac{P_{\text{total}}}{A_{\text{raft}}} - \gamma D_f \qquad \text{[U.S.]} \qquad 36.24(b)$$

CERM Chapter 37
Rigid Retaining Walls

3. EARTH PRESSURE

$$\alpha = 45° + \frac{\phi}{2} \quad [\text{Rankine}] \qquad 37.1$$

$$\alpha = \phi + \arctan\left(\frac{\begin{array}{c} -\tan\phi \\ + \sqrt{\begin{array}{c} \tan\phi(\tan\phi + \cot\phi) \\ \times (1 + \tan\delta\cot\phi) \end{array}} \end{array}}{1 + \tan\delta(\tan\phi + \cot\phi)} \right)$$

[Coulomb]

37.2

Figure 37.2 *Active and Passive Earth Pressure*

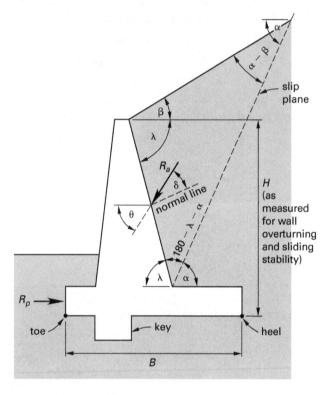

4. VERTICAL SOIL PRESSURE

$$p_v = \rho g H \qquad \text{[SI]} \qquad 37.3(a)$$

$$p_v = \gamma H \qquad \text{[U.S.]} \qquad 37.3(b)$$

5. ACTIVE EARTH PRESSURE

$$p_a = p_v k_a - 2c\sqrt{k_a} \qquad 37.4$$

$$k_a = \frac{\sin^2(\lambda + \phi)}{\sin^2\lambda \sin(\lambda - \delta)\left(1 + \sqrt{\dfrac{\sin(\phi + \delta)\sin(\phi - \beta)}{\sin(\lambda - \delta)\sin(\lambda + \beta)}}\right)^2}$$

[Coulomb]

37.5

$$k_a = \cos\beta\left(\frac{\cos\beta - \sqrt{\cos^2\beta - \cos^2\phi}}{\cos\beta + \sqrt{\cos^2\beta - \cos^2\phi}}\right)$$

[Rankine]

37.6

$$k_a = \frac{1}{k_p} = \tan^2\left(45° - \frac{\phi}{2}\right)$$

$$= \frac{1 - \sin\phi}{1 + \sin\phi} \quad \left[\begin{array}{l} \text{Rankine: horizontal} \\ \text{backfill; vertical face} \end{array}\right] \qquad 37.7$$

$$p_a = p_v - 2c \quad [\phi = 0°] \qquad 37.8$$

$$p_a = k_a p_v \quad [c = 0] \qquad 37.9$$

$$R_a = \tfrac{1}{2} p_a H = \tfrac{1}{2} k_a \rho g H^2 \qquad \text{[SI]} \qquad 37.10(a)$$

$$R_a = \tfrac{1}{2} p_a H = \tfrac{1}{2} k_a \gamma H^2 \qquad \text{[U.S.]} \qquad 37.10(b)$$

$$\theta_R = 90° \text{ from the wall} \quad \text{[Rankine]} \qquad 37.11$$

$$\theta_R = 90° - \delta \text{ from the wall} \quad \text{[Coulomb]} \qquad 37.12$$

6. PASSIVE EARTH PRESSURE

$$p_p = p_v k_p + 2c\sqrt{k_p} \qquad 37.13$$

$$k_p = \frac{\sin^2(\lambda - \phi)}{\sin^2\lambda \sin(\lambda + \delta)\left(1 - \sqrt{\dfrac{\sin(\phi + \delta)\sin(\phi + \beta)}{\sin(\lambda + \delta)\sin(\lambda + \beta)}}\right)^2}$$

[Coulomb]

37.14

$$k_p = \cos\beta\left(\frac{\cos\beta + \sqrt{\cos^2\beta - \cos^2\phi}}{\cos\beta - \sqrt{\cos^2\beta - \cos^2\phi}}\right) \quad \text{[Rankine]}$$

37.15

$$k_p = \frac{1}{k_a} = \tan^2\left(45° + \frac{\phi}{2}\right)$$

$$= \frac{1 + \sin\phi}{1 - \sin\phi} \quad \left[\begin{array}{l} \text{Rankine: horizontal} \\ \text{backfill; vertical face} \end{array}\right] \qquad 37.16$$

$$p_p = p_v + 2c \quad [\phi = 0°] \qquad 37.17$$

$$p_p = k_p p_v \quad [c = 0] \qquad 37.18$$

$$R_p = \tfrac{1}{2}p_p H = \tfrac{1}{2}k_p \rho g H^2 \qquad \text{[SI]} \qquad 37.19(a)$$

$$R_p = \tfrac{1}{2}p_p H = \tfrac{1}{2}k_p \gamma H^2 \qquad \text{[U.S.]} \qquad 37.19(b)$$

7. AT-REST SOIL PRESSURE

$$p_o = k_o p_v \qquad 37.20$$

$$k_o \approx 1 - \sin\phi \qquad 37.21$$

$$R_o = \tfrac{1}{2}k_o \rho g H^2 \qquad \text{[SI]} \qquad 37.22(a)$$

$$R_o = \tfrac{1}{2}k_o \gamma H^2 \qquad \text{[U.S.]} \qquad 37.22(b)$$

10. SURCHARGE LOADING

$$p_q = k_a q \qquad 37.31$$

$$R_q = k_a q H w \qquad 37.32$$

$$p_q = \frac{1.77 V_q m^2 n^2}{H^2(m^2 + n^2)^3} \qquad [m > 0.4] \qquad 37.33$$

$$p_q = \frac{0.28 V_q n^2}{H^2(0.16 + n^2)^3} \qquad [m \le 0.4] \qquad 37.34$$

$$R_q \approx \frac{0.78 V_q}{H} \qquad [m = 0.4] \qquad 37.35$$

$$R_q \approx \frac{0.60 V_q}{H} \qquad [m = 0.5] \qquad 37.36$$

$$R_q \approx \frac{0.46 V_q}{H} \qquad [m = 0.6] \qquad 37.37$$

$$m = \frac{x}{H} \qquad 37.38$$

$$n = \frac{y}{H} \qquad 37.39$$

$$p_q = \frac{4 L_q m^2 n}{\pi H(m^2 + n^2)^2} \qquad [m > 0.4] \qquad 37.40$$

$$R_q = \frac{0.64 L_q}{m^2 + 1} \qquad 37.41$$

$$p_q = \frac{0.203 L_q n}{H(0.16 + n^2)^2} \qquad [m \le 0.4] \qquad 37.42$$

$$R_q = 0.55 L_q \qquad 37.43$$

Figure 37.3 Surcharges

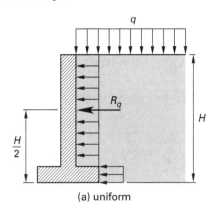

(a) uniform

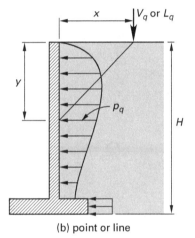

(b) point or line

11. EFFECTIVE STRESS

$$\mu = \rho_w g h \qquad \text{[SI]} \qquad 37.44(a)$$

$$\mu = \gamma_w h \qquad \text{[U.S.]} \qquad 37.44(b)$$

$$\rho_{\text{sat}} = \rho_{\text{dry}} + n\rho_w$$
$$= \rho_{\text{dry}} + \left(\frac{e}{1+e}\right)\rho_w \qquad \text{[SI]} \qquad 37.45(a)$$

$$\gamma_{\text{sat}} = \gamma_{\text{dry}} + n\gamma_w$$
$$= \gamma_{\text{dry}} + \left(\frac{e}{1+e}\right)\gamma_w \qquad \text{[U.S.]} \qquad 37.45(b)$$

$$p_v = g(\rho_{\text{sat}} H - \rho_w h) \qquad \text{[SI]} \qquad 37.46(a)$$

$$p_v = \gamma_{\text{sat}} H - \gamma_w h \qquad \text{[U.S.]} \qquad 37.46(b)$$

$$p_h = g\big(\rho_w h + k_a(\rho_{\text{sat}} H - \rho_w h)\big)$$
$$= g\big(k_a \rho_{\text{sat}} H + (1 - k_a)\rho_w h\big) \qquad \text{[SI]} \qquad 37.47(a)$$

$$p_h = \gamma_w h + k_a(\gamma_{\text{sat}} H - \gamma_w h)$$
$$= k_a \gamma_{\text{sat}} H + (1 - k_a)\gamma_w h \qquad \text{[U.S.]} \qquad 37.47(b)$$

$$\rho_{eq} = (1 - k_a)\rho_w \qquad \text{[SI]} \quad 37.48(a)$$

$$\gamma_{eq} = (1 - k_a)\gamma_w \qquad \text{[U.S.]} \quad 37.48(b)$$

12. CANTILEVER RETAINING WALLS: ANALYSIS

$$W_i = g\rho_i A_i \qquad \text{[SI]} \quad 37.49(a)$$

$$W_i = \gamma_i A_i \qquad \text{[U.S.]} \quad 37.49(b)$$

$$M_{\text{toe}} = \sum W_i x_i - R_{a,h} y_{a,h} + R_{a,v} x_{a,v} \qquad 37.50$$

$$x_R = \frac{M_{\text{toe}}}{R_{a,v} + \sum W_i} \qquad 37.51$$

$$\epsilon = \left| \frac{B}{2} - x_R \right| \qquad 37.52$$

$$F_{\text{OT}} = \frac{M_{\text{resisting}}}{M_{\text{overturning}}}$$

$$= \frac{\sum W_i x_i + R_{a,v} x_{a,v}}{R_{a,h} y_{a,h}} \qquad 37.53$$

$$p_{v,\max}, p_{v,\min} = \left(\frac{\sum W_i + R_{a,v}}{B} \right)$$

$$\times \left(1 \pm \frac{6\epsilon}{B} \right) \qquad 37.54$$

$$R_{\text{SL}} = \left(\sum W_i + R_{a,v} \right) \tan \phi + c_A B \qquad \text{[key]} \quad 37.55$$

$$R_{\text{SL}} = \left(\sum W_i + R_{a,v} \right) \tan \delta + c_A B \qquad \text{[no key]} \quad 37.56$$

$$F_{\text{SL}} = \frac{R_{\text{SL}}}{R_{a,h}} \qquad 37.57$$

Figure 37.4 Elements Contributing to Vertical Force* (step 3)

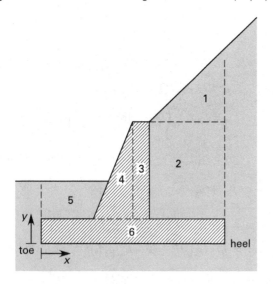

*Some retaining walls may not have all elements.

Figure 37.5 Resultant Distribution on the Base (step 7)

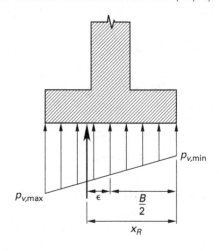

13. CANTILEVER RETAINING WALLS: DESIGN

$$R_{a,h}\left(\frac{H}{3} \right) \approx \left(W_{\text{above heel}} \right)\left(\frac{L}{2} \right) \qquad 37.58$$

$$B \approx \tfrac{3}{2}L \qquad 37.59$$

$$8 < \frac{H}{t_{\text{stem}}} < 12 \qquad 37.60$$

$$10 < \frac{H}{t_{\text{base}}} < 14 \qquad 37.61$$

CERM Chapter 38
Piles and Deep Foundations

1. INTRODUCTION

$$Q_{\text{ult}} = Q_p + Q_f \qquad 38.1$$

$$Q_a = \frac{Q_{\text{ult}}}{F} \qquad 38.2$$

2. PILE CAPACITY FROM DRIVING DATA

$$Q_{a,\text{lbf}} = \frac{Q_{\text{ult}}}{FS} = \frac{2 W_{\text{hammer,lbf}} H_{\text{fall,ft}}}{S_{\text{in}} + 1} \qquad \text{[drop hammer]}$$

$$38.3$$

$$Q_{a,\text{lbf}} = \frac{2 W_{\text{hammer,lbf}} H_{\text{fall,ft}}}{S_{\text{in}} + 0.1}$$

$$\begin{bmatrix} \text{single-acting steam hammer;} \\ \text{driven weight} < \text{striking weight} \end{bmatrix} \qquad 38.4$$

$$Q_{a,\text{lbf}} = \frac{2W_{\text{hammer,lbf}}H_{\text{fall,ft}}}{S_{\text{in}} + 0.1\left(\dfrac{W_{\text{driven}}}{W_{\text{hammer}}}\right)}$$

$$\left[\begin{array}{l}\text{single-acting steam hammer;}\\\text{driven weight} > \text{striking weight}\end{array}\right] \quad 38.5$$

$$Q_{a,\text{lbf}} = \frac{2E_{\text{ft-lbf}}}{S_{\text{in}} + 0.1}$$

$$\left[\begin{array}{l}\text{double-acting steam hammer;}\\\text{driven weight} < \text{striking weight}\end{array}\right] \quad 38.6$$

$$Q_{a,\text{lbf}} = \frac{2E_{\text{ft-lbf}}}{S_{\text{in}} + 0.1\left(\dfrac{W_{\text{driven}}}{W_{\text{hammer}}}\right)}$$

$$\left[\begin{array}{l}\text{double-acting steam hammer;}\\\text{driven weight} > \text{striking weight}\end{array}\right] \quad 38.7$$

3. THEORETICAL POINT-BEARING CAPACITY

$$Q_p = A_p\left(\tfrac{1}{2}\rho g B N_\gamma + cN_c + \rho g D_f N_q\right) \quad \text{[SI]} \quad 38.8(a)$$

$$Q_p = A_p\left(\tfrac{1}{2}\gamma B N_\gamma + cN_c + \gamma D_f N_q\right) \quad \text{[U.S.]} \quad 38.8(b)$$

$$Q_p = A_p\rho g D N_q \quad \text{[cohesionless; } D \leq D_c] \quad \text{[SI]} \quad 38.9(a)$$

$$Q_p = A_p\gamma D N_q \quad \text{[cohesionless; } D \leq D_c] \quad \text{[U.S.]} \quad 38.9(b)$$

$$Q_p = A_p c N_c \approx 9A_p c \quad \text{[cohesive]} \quad 38.10$$

4. THEORETICAL SKIN-FRICTION CAPACITY

$$Q_f = A_s f_s = p f_s L_e = p f_s(L - \text{seasonal variation})$$
$$38.11$$

$$Q_f = p\sum f_{s,i}L_{e,i} \quad 38.12$$

$$f_s = c_A + \sigma_h \tan\delta \quad 38.13$$

$$f_s = \alpha c \quad 38.14$$

$$\sigma_h = k_s\sigma'_v = k_s(\rho g D - \mu) \quad \text{[SI]} \quad 38.15(a)$$

$$\sigma_h = k_s\sigma'_v = k_s(\gamma D - \mu) \quad \text{[U.S.]} \quad 38.15(b)$$

$$\mu = \rho_w g h \quad \text{[SI]} \quad 38.16(a)$$

$$\mu = \gamma_w h \quad \text{[U.S.]} \quad 38.16(b)$$

$$Q_f = p k_s \tan\delta\sum L_i\sigma' \quad \text{[cohesionless]} \quad 38.17$$

$$Q_f = p\sum c_A L_i \quad \text{[cohesive]} \quad 38.18$$

$$Q_f = p\beta\sigma' L \quad \text{[cohesive]} \quad 38.19$$

7. CAPACITY OF PILE GROUPS

$$Q_s = 2(b + w)L_e c_1 \quad 38.20$$

$$Q_p = 9c_2 bw \quad 38.21$$

$$Q_{\text{ult}} = Q_s + Q_p \quad 38.22$$

$$Q_a = \frac{Q_{\text{ult}}}{F} \quad 38.23$$

$$\eta_G = \frac{Q_G}{\sum Q_i} \quad 38.24$$

CERM Chapter 39
Excavations

3. BRACED CUTS IN SAND

$$p_{\max} = 0.65k_a\rho gH \quad \text{[SI]} \quad 39.1(a)$$

$$p_{\max} = 0.65k_a\gamma H \quad \text{[U.S.]} \quad 39.1(b)$$

Figure 39.2 *Cuts in Sand*

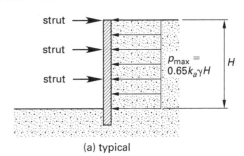

(a) typical

(b) Tschebotarioff

4. BRACED CUTS IN STIFF CLAY

$$0.2\rho gH \leq p_{\max} \leq 0.4\rho gH \quad \text{[SI]} \quad 39.2(a)$$

$$0.2\gamma H \leq p_{\max} \leq 0.4\gamma H \quad \text{[U.S.]} \quad 39.2(b)$$

$$p_{\max} = k_a \rho g H \quad \text{[SI]} \quad 39.3(a)$$

$$p_{\max} = k_a \gamma H \quad \text{[U.S.]} \quad 39.3(b)$$

$$k_a = 1 - \frac{4c}{\rho g H} \quad \text{[SI]} \quad 39.4(a)$$

$$k_a = 1 - \frac{4c}{\gamma H} \quad \text{[U.S.]} \quad 39.4(b)$$

Figure 39.3 *Cuts in Stiff Clay*

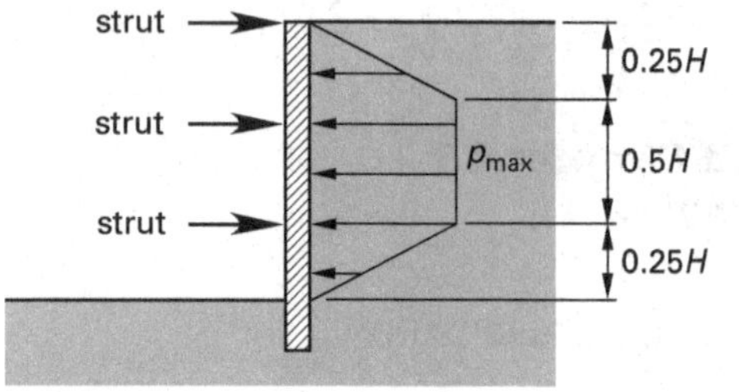

5. BRACED CUTS IN SOFT CLAY

$$p_{\max} = \rho g H - 4c \quad \text{[SI]} \quad 39.5(a)$$

$$p_{\max} = \gamma H - 4c \quad \text{[U.S.]} \quad 39.5(b)$$

Figure 39.4 *Cuts in Soft Clay*

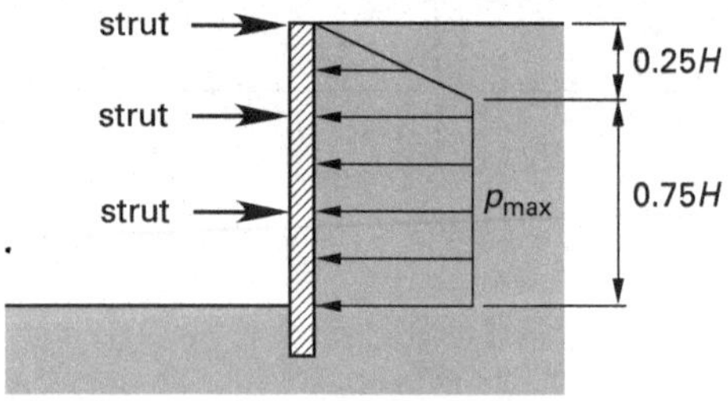

7. STABILITY OF BRACED EXCAVATIONS IN CLAY

$$H_c = \frac{5.7c}{\rho g - \sqrt{2}\left(\frac{c}{B}\right)} \quad [H < B] \quad \text{[SI]} \quad 39.6(a)$$

$$H_c = \frac{5.7c}{\gamma - \sqrt{2}\left(\frac{c}{B}\right)} \quad [H < B] \quad \text{[U.S.]} \quad 39.6(b)$$

$$H_c = \frac{N_c c}{\rho g} \quad [H > B] \quad \text{[SI]} \quad 39.7(a)$$

$$H_c = \frac{N_c c}{\gamma} \quad [H > B] \quad \text{[U.S.]} \quad 39.7(b)$$

$$F = \frac{N_c c}{\rho g H + q} \quad \text{[SI]} \quad 39.8(a)$$

$$F = \frac{N_c c}{\gamma H + q} \quad \text{[U.S.]} \quad 39.8(b)$$

8. STABILITY OF BRACED EXCAVATIONS IN SAND

$$F = 2 N_\gamma k_a \tan\phi \quad 39.9$$

$$F = 2 N_\gamma \left(\frac{\rho_{\text{submerged}}}{\rho_{\text{drained}}}\right) k_a \tan\phi \quad \text{[SI]} \quad 39.10(a)$$

$$F = 2 N_\gamma \left(\frac{\gamma_{\text{submerged}}}{\gamma_{\text{drained}}}\right) k_a \tan\phi \quad \text{[U.S.]} \quad 39.10(b)$$

10. ANALYSIS/DESIGN OF BRACED EXCAVATIONS

$$S = \frac{M_{\text{max,sheet piling}}}{F_b} \quad 39.11$$

11. ANALYSIS/DESIGN OF FLEXIBLE BULKHEADS

$$y = \frac{k_p D^2 - k_a (H + D)^2}{(k_p - k_a)(H + 2D)} \quad \text{[cohesionless]} \quad 39.12$$

Figure 39.5 *Net Horizontal Pressure Distribution on a Bulkhead in Uniform Granular Soil (simplified analysis)*

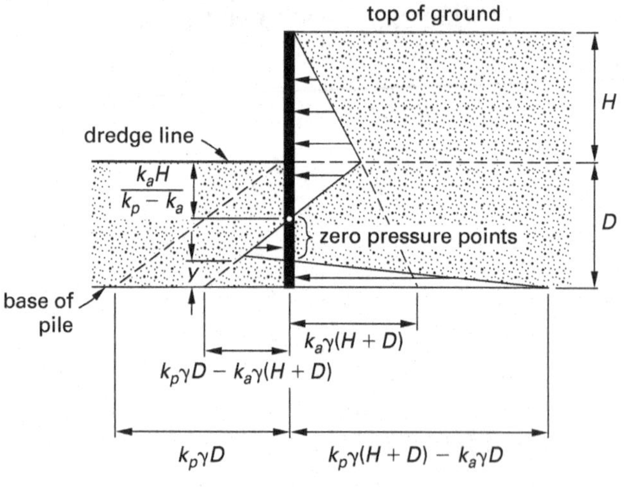

13. ANALYSIS/DESIGN OF ANCHORED BULKHEADS

$$f = \frac{M}{S} \quad 39.13$$

$$S = \frac{M}{F_b} \quad 39.14$$

CERM Chapter 40
Special Soil Topics

1. PRESSURE FROM APPLIED LOADS: BOUSSINESQ'S EQUATION

$$\Delta p_v = \frac{3h^3 P}{2\pi s^5}$$

$$= \left(\frac{3P}{2\pi h^2}\right)\left(\frac{1}{1 + \left(\frac{r}{h}\right)^2}\right)^{5/2} \quad [h > 2B] \qquad 40.1$$

Figure 40.1 Pressure at a Point

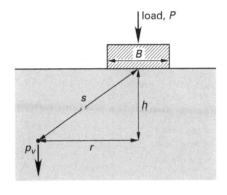

2. PRESSURE FROM APPLIED LOADS: ZONE OF INFLUENCE

$$A = \left(B + 2(h \cot 60°)\right)\left(L + 2(h \cot 60°)\right) \qquad 40.2$$

$$A = (B + h)(L + h) \qquad 40.3$$

$$\Delta p_v = \frac{P}{A} \qquad 40.4$$

3. PRESSURE FROM APPLIED LOADS: INFLUENCE CHART

$$\Delta p_v = I(\text{no. of squares})p_{\text{app}} \qquad 40.5$$

5. CLAY CONDITION

Figure 40.4 Consolidation Curve for Clay

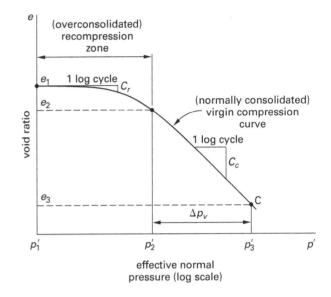

6. CONSOLIDATION PARAMETERS

$$C_r = \frac{-(e_1 - e_2)}{\log_{10}\frac{p_1'}{p_2'}} = \frac{e_2 - e_1}{\log_{10}\frac{p_1'}{p_2'}} \qquad 40.6$$

$$C_c = \frac{-(e_2 - e_3)}{\log_{10}\frac{p_2'}{p_3'}} = \frac{e_3 - e_2}{\log_{10}\frac{p_2'}{p_3'}}$$

$$= \frac{\Delta e}{\log_{10}\frac{p_2'}{p_2' + \Delta p_v'}} \qquad 40.7$$

$$C_c \approx 1.15(e_o - 0.35) \quad [\text{clays}] \qquad 40.8$$

- *normally consolidated clays*

$$C_c \approx 0.009(\text{LL} - 10) \qquad 40.9$$

- *organic soils*

$$C_c \approx 0.0155w \qquad 40.10$$

- *varved clays*

$$C_c \approx (1 + e_o)\left(0.1 + 0.006(w_n - 25)\right) \qquad 40.11$$

$$\text{CR} = \frac{C_c}{1 + e_o} \qquad 40.12$$

$$\text{RR} = \frac{C_r}{1 + e_o} \qquad 40.13$$

$$e_o = w_o(\text{SG}) \quad [\text{saturated}] \qquad 40.14$$

7. PRIMARY CONSOLIDATION

$$S_{\text{primary}} = \frac{H\Delta e}{1 + e_o}$$

$$= \frac{H C_r \log_{10} \dfrac{p'_o + \Delta p'_v}{p'_o}}{1 + e_o}$$

$$= H(\text{RR})\log_{10} \frac{p'_o + \Delta p'_v}{p'_o} \quad \text{[overconsolidated]}$$

40.15

$$S_{\text{primary}} = \frac{H\Delta e}{1 + e_o}$$

$$= \frac{H C_c \log_{10} \dfrac{p'_o + \Delta p'_v}{p'_o}}{1 + e_o}$$

$$= H(\text{CR})\log_{10} \frac{p'_o + \Delta p'_v}{p'_o} \quad \text{[normally consolidated]}$$

40.16

$$p_v = \gamma_{\text{layer}} H \quad \text{[above GWT]} \qquad 40.17$$

$$p'_v = \gamma_{\text{layer}} H - \gamma_{\text{water}} h$$

$$= p_v - \mu \quad \text{[below GWT]} \qquad 40.18$$

$$\mu = \gamma_{\text{water}} h \qquad 40.19$$

8. PRIMARY CONSOLIDATION RATE

$$t = \frac{T_v H_d^2}{C_v} \qquad 40.20$$

$$C_v = \frac{K(1 + e_o)}{a_v \gamma_{\text{water}}} \qquad 40.21$$

$$a_v = \frac{-(e_2 - e_1)}{p'_2 - p'_1} \qquad 40.22$$

$$T_v = \tfrac{1}{4}\pi U_z^2 \quad [U_z < 0.60] \qquad 40.23(a)$$

$$T_v = 1.781 - 0.933\log\Big(100(1 - U_z)\Big) \quad [U_z \geq 0.60]$$

40.23(b)

Figure 40.5 *Consolidation Parameters*

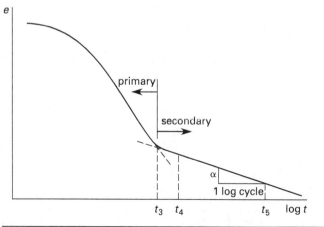

9. SECONDARY CONSOLIDATION

$$\alpha = \frac{-(e_5 - e_4)}{\log \dfrac{t_5}{t_4}} \qquad 40.24$$

$$C_\alpha = \frac{\alpha}{1 + e_o} \qquad 40.25$$

$$S_{\text{secondary}} = C_\alpha H \log_{10} \frac{t_5}{t_4} \qquad 40.26$$

10. SLOPE STABILITY IN SATURATED CLAY

$$d = \frac{D}{H} \qquad 40.27$$

$$F_{\text{cohesive}} = \frac{N_o c}{\gamma_{\text{eff}} H} \qquad 40.28$$

$$\gamma_{\text{eff}} = \gamma_{\text{saturated}} - \gamma_{\text{water}} \quad \text{[submerged]} \qquad 40.29$$

Figure 40.6 *Taylor Slope Stability (undrained, cohesive soils; $\phi = 0°$)*

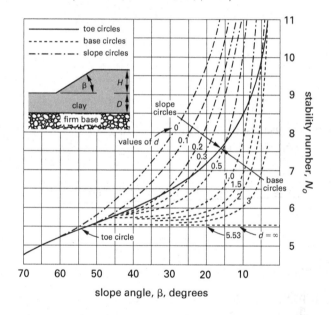

Source: *Soil Mechanics*, NAVFAC Design Manual DM-7.1, 1986, Fig. 2, p. 7.1-319.

11. LOADS ON BURIED PIPES

- *Marston's formula; rigid pipe*

$$w = C\rho g B^2 \qquad \text{[SI]} \quad 40.30(a)$$

$$w = C\gamma B^2 \qquad \text{[U.S.]} \quad 40.30(b)$$

$$p = \frac{w}{B} \qquad 40.31$$

- *flexible pipe*

$$w = C\rho g B D \qquad \text{[SI]} \qquad 40.32(a)$$

$$w = C\gamma B D \qquad \text{[U.S.]} \qquad 40.32(b)$$

- *broad fill*

$$w = C_p \rho g D^2 \qquad \text{[SI]} \qquad 40.33(a)$$

$$w = C_p \gamma D^2 \qquad \text{[U.S.]} \qquad 40.33(b)$$

$$B_{\text{transition}} = D\sqrt{\frac{C_p}{C}} \qquad 40.34$$

Figure 40.8 *Pipes in Backfilled Trenches*

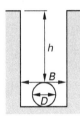 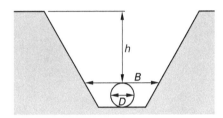

12. ALLOWABLE PIPE LOADS

$$\text{crushing strength} = (\text{D-load strength})D_i \qquad 40.35$$

$$w_{\text{allowable}} = \left(\frac{\text{known pipe}}{\text{crushing strength}} \right)\left(\frac{\text{LF}}{F}\right) \qquad \text{[analysis]}$$

$$40.36$$

17. LIQUEFACTION

$$\frac{\tau_{h,\text{ave}}}{\sigma'_o} \approx 0.65\left(\frac{a_{\max}}{g}\right)\left(\frac{\sigma_o}{\sigma'_o}\right)r_d \qquad 40.37$$

Structural

CERM Chapter 41
Determinate Statics

4. CONCENTRATED FORCES

$$F_x = F\cos\theta_x \qquad 41.3$$

$$F_y = F\cos\theta_y \qquad 41.4$$

$$F_z = F\cos\theta_z \qquad 41.5$$

$$F = \sqrt{F_x^2 + F_y^2 + F_z^2} \qquad 41.6$$

Figure 41.1 *Components and Direction Angles of a Force*

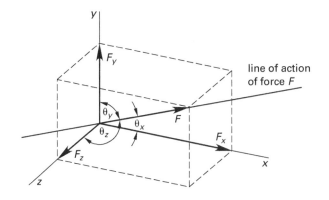

6. MOMENT OF A FORCE ABOUT A POINT

$$\mathbf{M}_O = \mathbf{r} \times \mathbf{F} \qquad 41.7$$

$$M_O = |\mathbf{M}_O| = |\mathbf{r}||\mathbf{F}|\sin\theta = d|\mathbf{F}| \quad [\theta \le 180°] \qquad 41.8$$

9. COMPONENTS OF A MOMENT

$$M_x = M\cos\theta_x \qquad 41.12$$

$$M_y = M\cos\theta_y \qquad 41.13$$

$$M_z = M\cos\theta_z \qquad 41.14$$

$$M_x = yF_z - zF_y \qquad 41.15$$

$$M_y = zF_x - xF_z \qquad 41.16$$

$$M_z = xF_y - yF_x \qquad 41.17$$

$$M = \sqrt{M_x^2 + M_y^2 + M_z^2} \qquad 41.18$$

10. COUPLES

$$M_O = 2rF\sin\theta = Fd \qquad 41.19$$

Figure 41.4 *Couple*

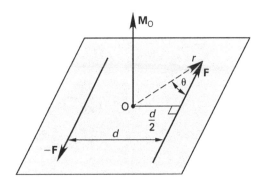

15. MOMENT FROM A DISTRIBUTED LOAD

$$M_{\text{distributed load}} = \text{force} \times \text{distance} = wx\left(\frac{x}{2}\right)$$

$$= \tfrac{1}{2}wx^2 \qquad 41.34$$

17. CONDITIONS OF EQUILIBRIUM

$$\mathbf{F}_R = \sum \mathbf{F} = 0 \qquad 41.35$$

$$F_R = \sqrt{F_{R,x}^2 + F_{R,y}^2 + F_{R,z}^2} = 0 \qquad 41.36$$

$$\mathbf{M}_R = \sum \mathbf{M} = 0 \qquad 41.37$$

$$M_R = \sqrt{M_{R,x}^2 + M_{R,y}^2 + M_{R,z}^2} = 0 \qquad 41.38$$

25. INFLUENCE LINES FOR REACTIONS

$$R = F \times \text{influence line ordinate} \qquad 41.45$$

27. LEVERS

$$\frac{\text{mechanical}}{\text{advantage}} = \frac{F_{\text{load}}}{F_{\text{applied}}} = \frac{\text{applied force lever arm}}{\text{load lever arm}}$$

$$= \frac{\text{distance moved by applied force}}{\text{distance moved by load}}$$

41.46

32. DETERMINATE TRUSSES

$$\text{no. of members} = 2(\text{no. of joints}) - 3 \qquad 41.48$$

no. of members

$+$ no. of reactions

$-\; 2(\text{no. of joints}) = 0 \quad \text{[determinate]}$

$> 0 \quad \text{[indeterminate]}$

$< 0 \quad \text{[unstable]} \qquad 41.49$

40. PARABOLIC CABLES

$$\sum M_{\text{D}} = wa\!\left(\frac{a}{2}\right) - HS = 0 \qquad 41.50$$

$$H = \frac{wa^2}{2S} \qquad 41.51$$

$$w = mg \qquad\qquad \text{[SI]} \quad 41.52(a)$$

$$w = \frac{mg}{g_c} \qquad\qquad \text{[U.S.]} \quad 41.52(b)$$

$$T_{\text{C},x} = H = \frac{wa^2}{2S} \qquad 41.53$$

$$T_{\text{C},y} = wx \qquad 41.54$$

$$T_{\text{C}} = \sqrt{T_{\text{C},x}^2 + T_{\text{C},y}^2}$$

$$= w\sqrt{\left(\frac{a^2}{2S}\right)^2 + x^2} \qquad 41.55$$

$$\tan\theta = \frac{wx}{H} \qquad 41.56$$

$$y(x) = \frac{wx^2}{2H} \qquad 41.57$$

Figure 41.18 *Parabolic Cable*

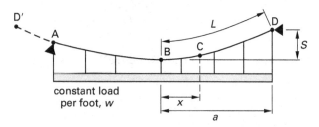

$$L \approx a\left(1 + \frac{2}{3}\!\left(\frac{S}{a}\right)^2 - \frac{2}{5}\!\left(\frac{S}{a}\right)^4\right) \qquad 41.58$$

42. CATENARY CABLES

Figure 41.20 *Catenary Cable*

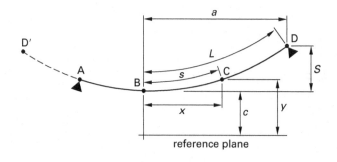

$$y(x) = c\cosh\frac{x}{c} \qquad 41.65$$

$$y = \sqrt{s^2 + c^2} = c\cosh\frac{x}{c} \qquad 41.66$$

$$s = c\sinh\frac{x}{c} \qquad 41.67$$

$$\text{sag} = S = y_{\text{D}} - c = c\left(\cosh\frac{a}{c} - 1\right) \qquad 41.68$$

$$\tan\theta = \frac{s}{c} \qquad 41.69$$

$$H = wc \qquad 41.70$$

$$F = ws \qquad 41.71$$

$$T = wy \qquad 41.72$$

$$\tan\theta = \frac{ws}{H} \qquad 41.73$$

$$\cos\theta = \frac{H}{T} \qquad 41.74$$

CERM Chapter 42
Properties of Areas

1. CENTROID OF AN AREA

(See the "Centroids and Area Moments of Inertia for Basic Shapes" table.)

$$x_c = \frac{\displaystyle\int x\,dA}{A} \qquad 42.1$$

$$y_c = \frac{\displaystyle\int y\,dA}{A} \qquad 42.2$$

$$A = \int f(x)\,dx \qquad 42.3$$

STRUCTURAL **65**

Centroids and Area Moments of Inertia for Basic Shapes

shape		centroidal location x_c	y_c	area, A	area moment of inertia (rectangular and polar), I, J	radius of gyration, r
rectangle		$\dfrac{b}{2}$	$\dfrac{h}{2}$	bh	$I_x = \dfrac{bh^3}{3}$ $I_{cx} = \dfrac{bh^3}{12}$ $J_c = \left(\dfrac{1}{12}\right)bh(b^2 + h^2)^*$ (see note below)	$r_x = \dfrac{h}{\sqrt{3}}$ $r_{cx} = \dfrac{h}{2\sqrt{3}}$
triangular area		$\dfrac{2b}{3}$	$\dfrac{h}{3}$	$\dfrac{bh}{2}$	$I_x = \dfrac{bh^3}{12}$ $I_{cx} = \dfrac{bh^3}{36}$	$r_x = \dfrac{h}{\sqrt{6}}$ $r_{cx} = \dfrac{h}{3\sqrt{2}}$
trapezoid		$\dfrac{2tz + t^2 + zb + tb + b^2}{3b + 3t}$	$\left(\dfrac{h}{3}\right)\left(\dfrac{b + 2t}{b + t}\right)$	$\dfrac{(b + t)h}{2}$	$I_x = \dfrac{(b + 3t)h^3}{12}$ $I_{cx} = \dfrac{(b^2 + 4bt + t^2)h^3}{36(b + t)}$	$r_x = \left(\dfrac{h}{\sqrt{6}}\right)\sqrt{\dfrac{b + 3t}{b + t}}$ $r_{cx} = \dfrac{h\sqrt{2(b^2 + 4bt + t^2)}}{6(b + t)}$
circle		0	0	πr^2	$I_x = I_y = \dfrac{\pi r^4}{4}$ $J_c = \dfrac{\pi r^4}{2}$	$r_x = \dfrac{r}{2}$
quarter-circular area		$\dfrac{4r}{3\pi}$	$\dfrac{4r}{3\pi}$	$\dfrac{\pi r^2}{4}$	$I_x = I_y = \dfrac{\pi r^4}{16}$ $J_o = \dfrac{\pi r^4}{8}$	
semicircular area		0	$\dfrac{4r}{3\pi}$	$\dfrac{\pi r^2}{2}$	$I_x = I_y = \dfrac{\pi r^4}{8}$ $I_{cx} = 0.1098r^4$ $J_o = \dfrac{\pi r^4}{4}$ $J_c = 0.5025r^4$	$r_x = \dfrac{r}{2}$ $r_{cx} = 0.264r$
quarter-elliptical area		$\dfrac{4a}{3\pi}$	$\dfrac{4b}{3\pi}$	$\dfrac{\pi ab}{4}$	$I_x = \dfrac{\pi ab^3}{8}$ $I_y = \dfrac{\pi a^3 b}{8}$	
semielliptical area		0	$\dfrac{4b}{3\pi}$	$\dfrac{\pi ab}{2}$	$J_o = \dfrac{\pi ab(a^2 + b^2)}{8}$	
semiparabolic area		$\dfrac{3a}{8}$	$\dfrac{3h}{5}$	$\dfrac{2ah}{3}$		
parabolic area		0	$\dfrac{3h}{5}$	$\dfrac{4ah}{3}$	$I_x = \dfrac{4ah^3}{7}$ $I_y = \dfrac{4ha^3}{15}$ $I_{cx} = \dfrac{16ah^3}{175}$	$r_x = h\sqrt{\dfrac{3}{7}}$ $r_y = \dfrac{a}{\sqrt{5}}$
parabolic spandrel		$\dfrac{3a}{4}$	$\dfrac{3h}{10}$	$\dfrac{ah}{3}$	$I_x = \dfrac{ah^3}{21}$ $I_y = \dfrac{3ha^3}{15}$	
general spandrel		$\left(\dfrac{n + 1}{n + 2}\right)a$	$\left(\dfrac{n + 1}{4n + 2}\right)h$	$\dfrac{ah}{n + 1}$	(note to accompany rectangular area above) *Theoretical definition based on $J = I_x + I_y$. However, in torsion, not all parts of the shape are effective. Effective values will be lower.	
circular sector [α in radians]		$\dfrac{2r\sin\alpha}{3\alpha}$	0	αr^2	$J = C\left(\dfrac{b^2 + h^2}{b^3 h^3}\right)$	

b/h	C
1	3.56
2	3.50
4	3.34
8	3.21

PPI • www.ppi2pass.com

$$x_c = \frac{\sum_i A_i x_{c,i}}{\sum_i A_i} \qquad 42.5$$

$$y_c = \frac{\sum_i A_i y_{c,i}}{\sum_i A_i} \qquad 42.6$$

2. FIRST MOMENT OF THE AREA

$$Q_y = \int x\,dA = x_c A \qquad 42.7$$

$$Q_x = \int y\,dA = y_c A \qquad 42.8$$

3. CENTROID OF A LINE

$$x_c = \frac{\int x\,dL}{L} \qquad 42.9$$

$$y_c = \frac{\int y\,dL}{L} \qquad 42.10$$

$$dL = \left(\sqrt{\left(\frac{dy}{dx}\right)^2 + 1}\right) dx \qquad 42.11$$

$$dL = \left(\sqrt{\left(\frac{dx}{dy}\right)^2 + 1}\right) dy \qquad 42.12$$

5. MOMENT OF INERTIA OF AN AREA

$$I_x = \int y^2\,dA \qquad 42.17$$

$$I_y = \int x^2\,dA \qquad 42.18$$

6. PARALLEL AXIS THEOREM

$$I_{\text{parallel axis}} = I_c + Ad^2 \qquad 42.20$$

7. POLAR MOMENT OF INERTIA

$$J = \int (x^2 + y^2)\,dA \qquad 42.21$$

$$J = I_x + I_y \qquad 42.22$$

$$J = I_{cx} + I_{cy} \qquad 42.23$$

8. RADIUS OF GYRATION

$$I = r^2 A \qquad 42.24$$

$$r = \sqrt{\frac{I}{A}} \qquad 42.25$$

$$r = \sqrt{\frac{J}{A}} \qquad 42.26$$

$$r^2 = r_x^2 + r_y^2 \qquad 42.27$$

10. SECTION MODULUS

$$S = \frac{I_c}{c} \qquad 42.30$$

CERM Chapter 43
Material Testing

2. TENSILE TEST

$$s = \frac{F}{A_o} \qquad 43.1$$

$$e = \frac{\delta}{L_o} \qquad 43.2$$

$$s = Ee \qquad 43.3$$

6. POISSON'S RATIO

$$\nu = \frac{e_{\text{lateral}}}{e_{\text{axial}}} = \frac{\dfrac{\Delta D}{D_o}}{\dfrac{\delta}{L_o}} \qquad 43.4$$

8. TRUE STRESS AND STRAIN

$$\sigma = \frac{F}{A} = \frac{F}{\left(1 - \dfrac{A_o - A}{A_o}\right)A_o} = \frac{F}{(1 - q)A_o}$$

$$= \frac{s}{1 - q}$$

$$= s(1 + e) \quad [\text{prior to necking, circular specimen}]$$
$$\qquad 43.5$$

$$q = \frac{A_o - A}{A_o}$$

$$\sigma = \frac{s}{(1 - \nu e)^2} \qquad 43.6$$

$$\epsilon = \int_{L_o}^{L} \frac{dL}{L} = \ln \frac{L}{L_o}$$

$$= \ln(1 + e) \quad \text{[prior to necking]} \qquad 43.7$$

$$A_o L_o = AL \qquad 43.8$$

$$\epsilon = \ln \frac{A_o}{A} = \ln \frac{D_o}{D})^2 = 2 \ln \frac{D_o}{D} \qquad 43.9$$

$$\sigma = K\epsilon^n \qquad 43.10$$

9. DUCTILITY

$$\text{percent elongation} = \frac{L_f - L_o}{L_o} \times 100\%$$

$$= e_f \times 100\% \qquad 43.11$$

$$\text{ductility} = \frac{\text{ultimate failure strain}}{\text{yielding strain}} \qquad 43.12$$

$$q_f = \frac{A_o - A_f}{A_o} \times 100\% \qquad 43.13$$

10. STRAIN ENERGY

$$\text{work} = \text{force} \times \text{distance} = \int F\,dL \qquad 43.14$$

$$\text{work per unit volume} = \int \frac{F\,dL}{AL} = \int_0^{\epsilon_\text{final}} \sigma\,d\epsilon \qquad 43.15$$

11. RESILIENCE

$$U_R = \int_0^{\epsilon_y} \sigma\,d\epsilon = E \int_0^{\epsilon_y} \epsilon\,d\epsilon = \frac{E\epsilon_y^2}{2}$$

$$= \frac{S_y \epsilon_y}{2} \qquad 43.16$$

12. TOUGHNESS

$$U_T \approx S_u \epsilon_u \quad \text{[ductile]} \qquad 43.17$$

$$U_T \approx \left(\frac{S_y + S_u}{2}\right)\epsilon_u \quad \text{[ductile]} \qquad 43.18$$

$$U_T \approx \tfrac{2}{3} S_u \epsilon_u \quad \text{[brittle]} \qquad 43.19$$

15. TORSION TEST

$$\tau = G\theta \qquad 43.21$$

$$G = \frac{E}{2(1 + \nu)} \qquad 43.22$$

$$\gamma = \frac{TL}{JG} = \frac{\tau L}{rG} \quad \text{[radians]} \qquad 43.23$$

$$S_{ys} = \frac{S_{yt}}{2} \quad \text{[maximum shear stress theory]} \qquad 43.24$$

$$S_{ys} = \frac{S_{yt}}{\sqrt{3}} = 0.577 S_{yt} \quad \text{[distortion energy theory]} \qquad 43.25$$

CERM Chapter 44
Strength of Materials

2. HOOKE'S LAW

$$\sigma = E\epsilon \qquad 44.2$$

$$\tau = G\phi \qquad 44.3$$

3. ELASTIC DEFORMATION

$$\delta = L_o\epsilon = \frac{L_o\sigma}{E} = \frac{L_o F}{EA} \qquad 44.4$$

$$L = L_o + \delta \qquad 44.5$$

4. TOTAL STRAIN ENERGY

$$U = \tfrac{1}{2}F\delta = \frac{F^2 L_o}{2AE} = \frac{\sigma^2 L_o A}{2E} \qquad 44.6$$

5. STIFFNESS AND RIGIDITY

$$k = \frac{F}{\delta} \quad \text{[general form]} \qquad 44.7(a)$$

$$k = \frac{AE}{L_o} \quad \text{[normal stress form]} \qquad 44.7(b)$$

$$R_j = \frac{k_j}{\sum_i k_i} \qquad 44.8$$

6. THERMAL DEFORMATION

$$\Delta L = \alpha L_o(T_2 - T_1) \qquad 44.9$$

$$\Delta A = \gamma A_o(T_2 - T_1) \qquad 44.10$$

$$\gamma \approx 2\alpha \qquad 44.11$$

$$\Delta V = \beta V_o (T_2 - T_1) \qquad 44.12$$

$$\beta \approx 3\alpha \qquad 44.13$$

$$\epsilon_{\text{th}} = \frac{\Delta L}{L_o} = \alpha (T_2 - T_1) \qquad 44.14$$

$$\sigma_{\text{th}} = E\epsilon_{\text{th}} \qquad 44.15$$

7. STRESS CONCENTRATIONS

$$\sigma' = K\sigma_0 \qquad 44.16$$

8. COMBINED STRESSES (BIAXIAL LOADING)

$$\sigma_\theta = \tfrac{1}{2}(\sigma_x + \sigma_y) + \tfrac{1}{2}(\sigma_x - \sigma_y)\cos 2\theta + \tau \sin 2\theta \qquad 44.17$$

$$\tau_\theta = -\tfrac{1}{2}(\sigma_x - \sigma_y)\sin 2\theta + \tau \cos 2\theta \qquad 44.18$$

$$\sigma_1, \sigma_2 = \tfrac{1}{2}(\sigma_x + \sigma_y) \pm \tau_1 \qquad 44.19$$

$$\tau_1, \tau_2 = \pm \tfrac{1}{2}\sqrt{(\sigma_x - \sigma_y)^2 + (2\tau)^2} \qquad 44.20$$

$$\theta_{\sigma_1,\sigma_2} = \tfrac{1}{2}\arctan\left(\frac{2\tau}{\sigma_x - \sigma_y}\right) \qquad 44.21$$

$$\theta_{\tau_1,\tau_2} = \tfrac{1}{2}\arctan\left(\frac{\sigma_x - \sigma_y}{-2\tau}\right) \qquad 44.22$$

Figure 44.6 Sign Convention for Combined Stress

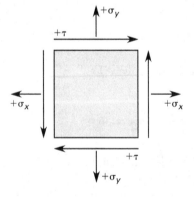

Figure 44.7 Stresses on an Inclined Plane

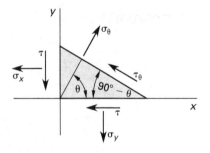

13. SHEAR STRESS IN BEAMS

$$\tau = \frac{V}{A} \qquad 44.28$$

$$\tau = \frac{V}{t_w d} \qquad 44.29$$

$$\tau_{y_1} = \frac{QV}{Ib} \qquad 44.30$$

$$Q = y^* A^* \qquad 44.32$$

$$\tau_{\text{max,rectangular}} = \frac{3V}{2A} = \frac{3V}{2bh} \qquad 44.33$$

$$\tau_{\text{max,circular}} = \frac{4V}{3A} = \frac{4V}{3\pi r^2} \qquad 44.34$$

$$\tau_{\text{max,hollow cylinder}} = \frac{2V}{A} \qquad 44.35$$

Figure 44.10 Web of a Flanged Beam

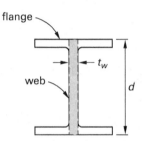

Figure 44.11 Shear Stress Distribution Within a Rectangular Beam

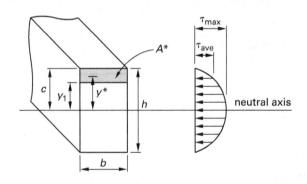

14. BENDING STRESS IN BEAMS

$$\sigma_b = \frac{-My}{I_c} \qquad 44.36$$

$$\sigma_{b,\text{max}} = \frac{Mc}{I_c} \qquad 44.37$$

$$\sigma_{b,\text{max}} = \frac{M}{S} \qquad 44.38$$

$$S = \frac{I_c}{c} \qquad 44.39$$

$$S_{\text{rectangular}} = \frac{bh^2}{6} \qquad 44.40$$

Figure 44.13 *Bending Stress Distribution in a Beam*

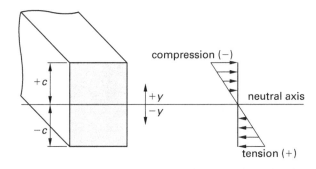

15. STRAIN ENERGY DUE TO BENDING MOMENT

$$U = \frac{1}{2EI} \int M^2(x)\,dx \qquad\qquad 44.41$$

16. ECCENTRIC LOADING OF AXIAL MEMBERS

$$\sigma_{\max,\min} = \frac{F}{A} \pm \frac{Mc}{I_c} \qquad\qquad 44.42$$

$$\sigma_{\max,\min} = \frac{F}{A} \pm \frac{Fec}{I_c} \qquad\qquad 44.43$$

Figure 44.14 *Eccentric Loading of an Axial Member*

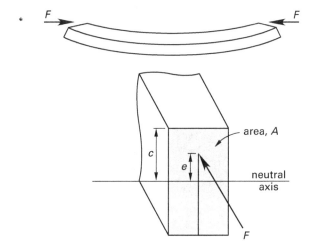

17. BEAM DEFLECTION: DOUBLE INTEGRATION METHOD

$$y = \text{deflection} \qquad\qquad 44.44$$

$$y' = \frac{dy}{dx} = \text{slope} \qquad\qquad 44.45$$

$$y'' = \frac{d^2y}{dx^2} = \frac{M(x)}{EI} \qquad\qquad 44.46$$

$$y''' = \frac{d^3y}{dx^3} = \frac{V(x)}{EI} \qquad\qquad 44.47$$

$$y = \frac{1}{EI} \int \left(\int M(x)\,dx \right) dx \qquad\qquad 44.48$$

18. BEAM DEFLECTION: MOMENT AREA METHOD

$$\phi = \int \frac{M(x)\,dx}{EI} \qquad\qquad 44.49$$

$$y = \int \frac{xM(x)\,dx}{EI} \qquad\qquad 44.50$$

19. BEAM DEFLECTION: STRAIN ENERGY METHOD

$$\tfrac{1}{2}Fy = \sum U \qquad\qquad 44.51$$

23. INFLECTION POINTS

$$y''(x) = \frac{1}{\rho(x)} = \frac{M(x)}{EI} \qquad\qquad 44.52$$

25. TRUSS DEFLECTION: VIRTUAL WORK METHOD

$$f\delta = \sum \frac{SuL}{AE} \quad [f=1] \qquad\qquad 44.53$$

28. COMPOSITE STRUCTURES

$$n = \frac{E}{E_{\text{weakest}}} \qquad\qquad 44.55$$

$$\sigma_{\text{weakest}} = \frac{F}{A_t} \qquad\qquad 44.56$$

$$\sigma_{\text{stronger}} = \frac{nF}{A_t} \qquad\qquad 44.57$$

$$\sigma_{\text{weakest}} = \frac{Mc_{\text{weakest}}}{I_{c,t}} \qquad\qquad 44.58$$

$$\sigma_{\text{stronger}} = \frac{nMc_{\text{stronger}}}{I_{c,t}} \qquad\qquad 44.59$$

CERM Chapter 45
Basic Elements of Design

1. ALLOWABLE STRESS DESIGN

$$\sigma_a = \frac{S_y}{\text{FS}} \quad [\text{ductile}] \qquad\qquad 45.2$$

$$\sigma_a = \frac{S_u}{\text{FS}} \quad [\text{brittle}] \qquad\qquad 45.3$$

2. ULTIMATE STRENGTH DESIGN

$$M_u = 1.2M_{\text{dead load}} + 1.6M_{\text{live load}} \qquad 45.5$$

$$M_n \geq \frac{M_u}{\phi} \qquad 45.6$$

3. SLENDER COLUMNS

r is the radius of gyration.

$$F_e = \frac{\pi^2 EI}{L^2} = \frac{\pi^2 EA}{\left(\dfrac{L}{r}\right)^2} \qquad 45.7$$

$$\sigma_e = \frac{F_e}{A} = \frac{\pi^2 E}{\left(\dfrac{L}{r}\right)^2} \quad [\sigma_e < \tfrac{1}{2}S_y] \qquad 45.8$$

$$L' = KL \qquad 45.9$$

Table 45.1 *Theoretical End Restraint Coefficients, K*

illus.	end conditions	ideal	recommended for design
(a)	both ends pinned	1	1.0[*]
(b)	both ends built in	0.5	0.65[*]–0.90
(c)	one end pinned, one end built in	0.707	0.80[*]–0.90
(d)	one end built in, one end free	2	2.0–2.1[*]
(e)	one end built in, one end fixed against rotation but free	1	1.2[*]
(f)	one end pinned, one end fixed against rotation but free	2	2.0[*]

[*]AISC values

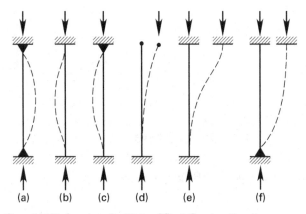

(a) (b) (c) (d) (e) (f)

$$\sigma_e = \frac{F_e}{A} = \frac{\pi^2 E}{\left(\dfrac{L'}{r}\right)^2} \quad [\sigma_e < \tfrac{1}{2}S_y] \qquad 45.10$$

$$(\text{SR})_T = \sqrt{\frac{2\pi^2 E}{S_y}} \qquad 45.11$$

Figure 45.1 *Euler's Curve*

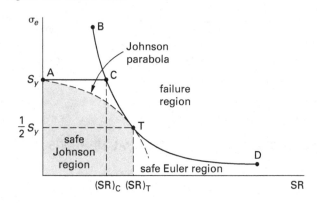

4. INTERMEDIATE COLUMNS

$$\sigma_{\text{cr}} = \frac{F_{\text{cr}}}{A} = a - b\left(\frac{KL}{r}\right)^2 \qquad 45.12$$

$$\sigma_{\text{cr}} = S_y - \left(\frac{1}{E}\right)\left(\frac{S_y}{2\pi}\right)^2\left(\frac{KL}{r}\right)^2 \qquad 45.13$$

5. ECCENTRICALLY LOADED COLUMNS

$$\sigma_{\max} = \sigma_{\text{ave}}(1 + \text{amplification factor})$$

$$= \left(\frac{F}{A}\right)\left(1 + \left(\frac{ec}{r^2}\right)\sec\left(\frac{\pi}{2}\sqrt{\frac{F}{F_e}}\right)\right)$$

$$= \left(\frac{F}{A}\right)\left(1 + \left(\frac{ec}{r^2}\right)\sec\left(\frac{L}{2r}\sqrt{\frac{F}{AE}}\right)\right)$$

$$= \left(\frac{F}{A}\right)\left(1 + \left(\frac{ec}{r^2}\right)\sec\phi\right) \qquad 45.14$$

$$\phi = \tfrac{1}{2}\left(\frac{L}{r}\right)\sqrt{\frac{F}{AE}} \qquad 45.15$$

6. THIN-WALLED CYLINDRICAL TANKS

$$\frac{t}{d_i} = \frac{t}{2r_i} < 0.1 \quad [\text{thin-walled}] \qquad 45.16$$

$$\sigma_h = \frac{pr}{t} \qquad 45.17$$

$$\sigma_l = \frac{pr}{2t} \qquad 45.18$$

$$\Delta L = L\epsilon_l = L\left(\frac{\sigma_l - \nu\sigma_h}{E}\right) \qquad 45.19$$

$$\Delta C = C\epsilon_h = \pi d_o\left(\frac{\sigma_h - \nu\sigma_l}{E}\right) \qquad 45.20$$

Figure 45.2 *Stresses in a Thin-Walled Tank*

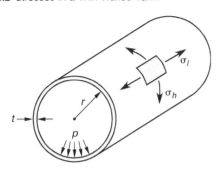

7. THICK-WALLED CYLINDERS

(See Table 45.2.)

$$\sigma_c = \frac{r_i^2 p_i - r_o^2 p_o + \dfrac{(p_i - p_o) r_i^2 r_o^2}{r^2}}{r_o^2 - r_i^2} \qquad 45.21$$

$$\sigma_r = \frac{r_i^2 p_i - r_o^2 p_o - \dfrac{(p_i - p_o) r_i^2 r_o^2}{r^2}}{r_o^2 - r_i^2} \qquad 45.22$$

$$\sigma_l = \frac{p_i r_i^2}{r_o^2 - r_i^2} \quad \begin{bmatrix} p_o \text{ does not act} \\ \text{longitudinally on the ends} \end{bmatrix} \qquad 45.23$$

$$\epsilon = \frac{\Delta d}{d} = \frac{\Delta C}{C} = \frac{\Delta r}{r}$$

$$= \frac{\sigma_c - \nu(\sigma_r + \sigma_l)}{E} \qquad 45.24$$

Figure 45.3 *Thick-Walled Cylinder*

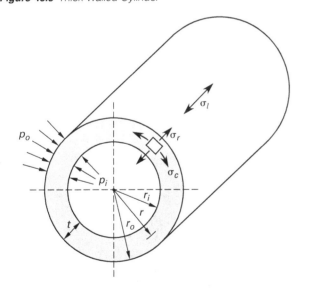

8. THIN-WALLED SPHERICAL TANKS

$$\sigma = \frac{pr}{2t} \qquad 45.25$$

Table 45.2 *Stresses in Thick-Walled Cylinders**

stress	external pressure, p	internal pressure, p
$\sigma_{c,o}$	$\dfrac{-(r_o^2 + r_i^2) p_o}{r_o^2 - r_i^2}$	$\dfrac{2 r_i^2 p_i}{r_o^2 - r_i^2}$
$\sigma_{r,o}$	$-p_o$	0
$\sigma_{c,i}$	$\dfrac{-2 r_o^2 p_o}{r_o^2 - r_i^2}$	$\dfrac{(r_o^2 + r_i^2) p_i}{r_o^2 - r_i^2}$
$\sigma_{r,i}$	0	$-p_i$
$\tau_{\max}$	$\frac{1}{2}\sigma_{c,i}$	$\frac{1}{2}(\sigma_{c,i} + p_i)$

*This table can be used with thin-walled cylinders. However, in most cases it will not be necessary to do so.

9. INTERFERENCE FITS

$$I_{\text{diametral}} = 2 I_{\text{radial}} = d_{o,\text{inner}} - d_{i,\text{outer}}$$

$$= |\Delta d_{o,\text{inner}}| + |\Delta d_{i,\text{outer}}| \qquad 45.26$$

$$\epsilon = \frac{\Delta d}{d} = \frac{\Delta C}{C} = \frac{\Delta r}{r}$$

$$= \frac{\sigma_c - \nu \sigma_r}{E} \qquad 45.27$$

$$I_{\text{diametral}} = 2 I_{\text{radial}}$$

$$= \left(\frac{2 p r_{o,\text{shaft}}}{E_{\text{hub}}} \right) \left(\frac{r_{o,\text{hub}}^2 + r_{o,\text{shaft}}^2}{r_{o,\text{hub}}^2 - r_{o,\text{shaft}}^2} + \nu_{\text{hub}} \right)$$

$$+ \left(\frac{2 p r_{o,\text{shaft}}}{E_{\text{shaft}}} \right) \left(\frac{r_{o,\text{shaft}}^2 + r_{i,\text{shaft}}^2}{r_{o,\text{shaft}}^2 - r_{i,\text{shaft}}^2} - \nu_{\text{shaft}} \right)$$

$$45.28$$

$$I_{\text{diametral}} = 2 I_{\text{radial}}$$

$$= \left(\frac{4 p r_{\text{shaft}}}{E} \right) \left(\frac{1}{1 - \left(\dfrac{r_{\text{shaft}}}{r_{o,\text{hub}}} \right)^2} \right) \qquad 45.29$$

$$F_{\max} = fN = 2\pi f p r_{o,\text{shaft}} L_{\text{interface}} \qquad 45.30$$

$$T_{\max} = 2\pi f p r_{o,\text{shaft}}^2 L_{\text{interface}} \qquad 45.31$$

12. RIVET AND BOLT CONNECTIONS

$$\tau = \frac{F}{A} = \frac{F}{\dfrac{\pi d^2}{4}} \qquad 45.32$$

$$n = \frac{\tau}{\text{allowable shear stress}} \qquad 45.33$$

$$A_t = t(b - nd) \qquad 45.34$$

$$\sigma_t = \frac{F}{A_t} \qquad 45.35$$

$$\sigma_p = \frac{F}{dt} \qquad 45.36$$

$$n = \frac{\sigma_p}{\sigma_{a,\text{bearing}}} \qquad 45.37$$

$$\tau = \frac{F}{2A} = \frac{F}{2dL} \qquad 45.38$$

13. BOLT PRELOAD

$$k_{\text{bolt}} = \frac{F}{\Delta L} = \frac{A_{\text{bolt}} E_{\text{bolt}}}{L} \qquad 45.39$$

$$k_{\text{parts}} = \frac{A_{e,\text{parts}} E_{\text{parts}}}{L} \qquad 45.40$$

$$\frac{1}{k_{\text{parts,composite}}} = \frac{1}{k_1} + \frac{1}{k_2} + \frac{1}{k_3} + \cdots \qquad 45.41$$

$$F_{\text{bolt}} = F_i + \frac{k_{\text{bolt}} F_{\text{applied}}}{k_{\text{bolt}} + k_{\text{parts}}} \qquad 45.42$$

$$F_{\text{parts}} = \frac{k_{\text{parts}} F_{\text{applied}}}{k_{\text{bolt}} + k_{\text{parts}}} - F_i \qquad 45.43$$

$$\sigma_{\text{bolt}} = \frac{KF}{A} \qquad 45.44$$

14. BOLT TORQUE TO OBTAIN PRELOAD

$$T = K_T d_{\text{bolt}} F_i \qquad 45.45$$

$$K_T = \frac{f_c r_c}{d_{\text{bolt}}} + \left(\frac{r_t}{d_{\text{bolt}}}\right)\left(\frac{\tan\theta + f_t \sec\alpha}{1 - f_t \tan\theta \sec\alpha}\right) \qquad 45.46$$

$$\tan\theta = \frac{\text{lead per revolution}}{2\pi r_t} \qquad 45.47$$

15. FILLET WELDS

$$t_e = \frac{\sqrt{2}}{2} y \qquad 45.48$$

$$\tau = \frac{F}{bt_e} \qquad 45.49$$

16. CIRCULAR SHAFT DESIGN

$$\tau = G\theta = \frac{Tr}{J} \qquad 45.50$$

$$U = \frac{T^2 L}{2GJ} \qquad 45.51$$

$$J = \frac{\pi r^4}{2} = \frac{\pi d^4}{32} \quad [\text{solid}] \qquad 45.52$$

$$J = \frac{\pi}{2}\left(r_o^4 - r_i^4\right) \quad [\text{hollow}] \qquad 45.53$$

$$\gamma = \frac{L\theta}{r} = \frac{TL}{GJ} \qquad 45.54$$

$$G = \frac{E}{2(1+\nu)} \qquad 45.55$$

$$T_{\text{N·m}} = \frac{9549 P_{\text{kW}}}{n_{\text{rpm}}} \qquad [\text{SI}] \quad 45.56(a)$$

$$T_{\text{in-lbf}} = \frac{63{,}025 P_{\text{horsepower}}}{n_{\text{rpm}}} \qquad [\text{U.S.}] \quad 45.56(b)$$

$$\tau_{\max} = \sqrt{\left(\frac{\sigma_x}{2}\right)^2 + \tau^2} \qquad 45.57$$

$$\tau_{\max} = \frac{16}{\pi d^3}\sqrt{M^2 + T^2} \qquad 45.58$$

$$\sigma' = \frac{16}{\pi d^3}\sqrt{4M^2 + 3T^2} \qquad 45.59$$

17. TORSION IN THIN-WALLED, NONCIRCULAR SHELLS

$$\tau = \frac{T}{2At} \qquad 45.60$$

$$q = \tau t = \frac{T}{2A} \quad [\text{constant}] \qquad 45.61$$

$$\gamma = \frac{TLp}{4A^2 tG} \qquad 45.62$$

20. ECCENTRICALLY LOADED BOLTED CONNECTIONS

$$\tau = \frac{Tr}{J} = \frac{Fer}{J} \qquad 45.63$$

$$J = \sum_i r_i^2 A_i \qquad 45.64$$

$$\tau_v = \frac{F}{nA} \qquad 45.65$$

23. SPRINGS

$$F = k\delta \qquad 45.66$$

$$k = \frac{F_1 - F_2}{\delta_1 - \delta_2} \qquad 45.67$$

$$\Delta E_p = \tfrac{1}{2}k\delta^2 \qquad 45.68$$

$$mg(h+\delta) = \tfrac{1}{2}k\delta^2 \qquad [\text{SI}] \quad 45.69(a)$$

$$m\left(\frac{g}{g_c}\right)(h+\delta) = \tfrac{1}{2}k\delta^2 \qquad [\text{U.S.}] \quad 45.69(b)$$

$$\frac{1}{k_{\text{eq}}} = \frac{1}{k_1} + \frac{1}{k_2} + \frac{1}{k_3} + \cdots \quad \begin{bmatrix} \text{series} \\ \text{springs} \end{bmatrix} \qquad 45.70$$

$$k_{\text{eq}} = k_1 + k_2 + k_3 + \cdots \quad \begin{bmatrix} \text{parallel} \\ \text{springs} \end{bmatrix} \qquad 45.71$$

24. WIRE ROPE

$$\sigma_b = \frac{d_w E_w}{d_{\text{sh}}} \qquad 45.72$$

$$F_a = \sigma_{a,\text{bending}} \times \text{number of strands} \times A_{\text{strand}} \qquad 45.73$$

$$p_p = \frac{2F_t}{d_r d_{\text{sh}}} \qquad 45.74$$

CERM Chapter 46
Structural Analysis I

2. DEGREE OF INDETERMINACY

For a two-dimensional truss,

$$I = r + m - 2j \qquad 46.1$$

For a three-dimensional truss,

$$I = r + m - 3j \qquad 46.2$$

For a rigid frame,

$$I = r + 3m - 3j - s \qquad 46.3$$

4. REVIEW OF ELASTIC DEFORMATION

This subject is covered in greater detail in CERM Chap. 44.

$$\delta = \frac{FL}{AE} \qquad 46.4$$

$$\delta = \alpha L_o \Delta T \qquad 46.5$$

7. THREE-MOMENT EQUATION

$$\frac{M_k L_k}{I_k} + 2M_{k+1}\left(\frac{L_k}{I_k} + \frac{L_{k+1}}{I_{k+1}}\right) + \frac{M_{k+2}L_{k+1}}{I_{k+1}}$$
$$= -6\left(\frac{A_k a}{I_k L_k} + \frac{A_{k+1} b}{I_{k+1} L_{k+1}}\right) \qquad 46.6$$

Figure 46.2 Portion of a Continuous Beam

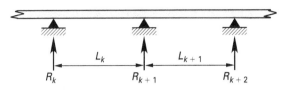

8. FIXED-END MOMENTS

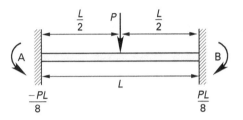

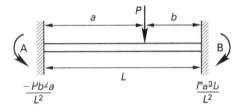

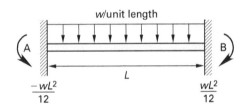

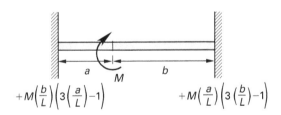

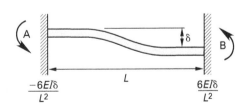

10. INFLUENCE DIAGRAMS

$$R_L = P \times \text{ordinate} \qquad \textit{46.11}$$

$$R_L = \int_{x_1}^{x_2} (w \times \text{ordinate}) \, dx$$

$$= w \times \text{area under curve} \qquad \textit{46.12}$$

CERM Chapter 47
Structural Analysis II

3. REVIEW OF WORK AND ENERGY

$$W = P\Delta \quad \text{[linear displacement]} \qquad \textit{47.1}$$

$$W = T\theta \quad \text{[rotation]} \qquad \textit{47.2}$$

$$W = U_2 - U_1 \qquad \textit{47.3}$$

4. REVIEW OF LINEAR DEFORMATION

$$\delta = \frac{PL}{AE} \qquad \textit{47.4}$$

5. THERMAL LOADING

$$P_{\text{th}} = \frac{\delta_{\text{constrained}} AE}{L} = \alpha(T_2 - T_1)\left(\frac{LAE}{L}\right)$$

$$= \alpha(T_2 - T_1)AE \qquad \textit{47.5}$$

$$M = \alpha(T_{\text{extreme fiber}} - T_{\text{neutral axis}})\left(\frac{EI}{\frac{h}{2}}\right) \qquad \textit{47.6}$$

6. DUMMY UNIT LOAD METHOD

$$W_Q = 1 \times \Delta = \int \sigma_Q \epsilon_P \, dV \qquad \textit{47.7}$$

$$W_Q = W_m + W_v + W_a + W_t \qquad \textit{47.8}$$

$$W_m = \int \frac{m_Q m_P}{EI} \, ds \qquad \textit{47.9}$$

$$W_v = \int \frac{V_Q V_P}{GA} \, ds \qquad \textit{47.10}$$

$$W_a = \int \frac{N_Q N_P}{EA} \, ds \qquad \textit{47.11}$$

$$W_t = \int \frac{T_Q T_P}{GJ} \, ds \qquad \textit{47.12}$$

7. BEAM DEFLECTIONS BY THE DUMMY UNIT LOAD METHOD

$$W_Q = W_m = \int \frac{m_Q m_P}{EI} \, ds \qquad \textit{47.13}$$

23. APPROXIMATE METHOD: MOMENT COEFFICIENTS

$$M = C_1 w L^2 \qquad \textit{47.43}$$

Table 47.2 ACI Moment Coefficients*

condition	C_1
positive moments near midspan	
end spans	
simply supported	$\frac{1}{11}$
built-in support	$\frac{1}{14}$
interior spans	$\frac{1}{16}$
negative moments at exterior face of first interior support	
2 spans	$\frac{1}{9}$
3 or more spans	$\frac{1}{10}$
negative moments at other faces or interior supports	
all cases	$\frac{1}{11}$
negative moments at face of all supports	
slabs with spans not exceeding 10 ft, and beams with ratio of sum of column stiffnesses to beam stiffness exceed 8 at each end of the span	$\frac{1}{12}$
negative moments at exterior built-in support	
support is a cross beam or girder (spandrel beam)	$\frac{1}{24}$
support is a column	$\frac{1}{16}$

*ACI 318 Sec. 6.5.2

24. APPROXIMATE METHOD: SHEAR COEFFICIENTS

$$V = C_2\left(\frac{wL}{2}\right) \qquad \textit{47.44}$$

C_2 has a value of 1.15 for end members at the first interior support. For shear at the face of all other supports, $C_2 = 1$.

CERM Chapter 48
Properties of Concrete and Reinforcing Steel

8. COMPRESSIVE STRENGTH

$$f_c' = \frac{P}{A} \qquad \textit{48.1}$$

10. MODULUS OF ELASTICITY

$$E_c = 0.043 w_c^{1.5} \sqrt{f_c'} \qquad \text{[SI]} \quad 48.2(a)$$

$$E_c = 33 w_c^{1.5} \sqrt{f_c'} \qquad \text{[U.S.]} \quad 48.2(b)$$

For normalweight concrete,

$$E_c = 5000 \sqrt{f_c'} \qquad \text{[SI]} \quad 48.3(a)$$

$$E_c = 57,000 \sqrt{f_c'} \qquad \text{[U.S.]} \quad 48.3(b)$$

11. SPLITTING TENSILE STRENGTH

$$f_{ct} = \frac{2P}{\pi DL} \qquad\qquad 48.4$$

$$f_{ct,\text{MPa}} = 0.56 \lambda \sqrt{f_{c,\text{MPa}}'} \qquad \text{[SI]} \quad 48.5(a)$$

$$f_{ct} = 6.7 \lambda \sqrt{f_c'} \qquad \text{[U.S.]} \quad 48.5(b)$$

Table 48.2 *Lightweight Aggregate Factors,* λ

normalweight concrete	1.0
sand-lightweight concrete	0.85
all-lightweight concrete	0.75

12. MODULUS OF RUPTURE

$$f_r = \frac{Mc}{I} \quad \text{[tension]} \qquad 48.6$$

$$f_r = 0.62 \lambda \sqrt{f_c'} \qquad \text{[SI]} \quad 48.7(a)$$

$$f_r = 7.5 \lambda \sqrt{f_c'} \qquad \text{[U.S.]} \quad 48.7(b)$$

CERM Chapter 49
Concrete Proportioning, Mixing, and Placing

6. ABSOLUTE VOLUME METHOD

$$V_{\text{absolute}} = \frac{m}{(\text{SG}) \rho_{\text{water}}} \qquad \text{[SI]} \quad 49.1(a)$$

$$V_{\text{absolute}} = \frac{W}{(\text{SG}) \gamma_{\text{water}}} \qquad \text{[U.S.]} \quad 49.1(b)$$

Table 49.2 *Summary of Approximate Properties of Concrete Components*

cement	
specific weight	195 lbf/ft^3 (3120 kg/m^3)
specific gravity	3.13–3.15
weight of one sack	94 lbf (42 kg)
fine aggregate	
specific weight	165 lbf/ft^3 (2640 kg/m^3)
specific gravity	2.64
coarse aggregate	
specific weight	165 lbf/ft^3 (2640 kg/m^3)
specific gravity	2.64
water	
specific weight	62.4 lbf/ft^3 (1000 kg/m^3)
	7.48 gal/ft^3 (1000 L/m^3)
	8.34 lbf/gal (1 kg/L)
	239.7 gal/ton (1 L/kg)
specific gravity	1.00

(Multiply lbf/ft^3 by 16.018 to obtain kg/m^3.)
(Multiply lbf by 0.4536 to obtain kg.)

7. ADJUSTMENTS FOR WATER AND AIR

(See Table 49.3.)

16. LATERAL PRESSURE ON FORMWORK

$$p = \rho g h \qquad \text{[SI]} \quad 49.2(a)$$

$$p = \gamma h = \frac{\rho g h}{g_c} \qquad \text{[U.S.]} \quad 49.2(b)$$

$$p_{\text{max,kPa}} = 30 C_w \le C_w C_c \left(7.2 + \frac{785 R_{\text{m/h}}}{T_{\circ\text{C}} + 17.8^\circ} \right) \le \rho g h$$

$$\begin{bmatrix} \text{columns: } h \le 1.2 \text{ m} \\ \text{walls: } h \le 4.2 \text{ m} \\ R \le 2.1 \text{ m/h} \end{bmatrix} \quad \text{[SI]}$$

$$49.3(a)$$

$$p_{\text{max,psf}} = 600 C_w \le C_w C_c \left(150 + \frac{9000 R_{\text{ft/hr}}}{T_{\circ\text{F}}} \right) \le \gamma h$$

$$\begin{bmatrix} \text{columns: } h \le 4 \text{ ft} \\ \text{walls: } h \le 14 \text{ ft} \\ R \le 7 \text{ ft/hr} \end{bmatrix} \quad \text{[U.S.]}$$

$$49.3(b)$$

Table 49.3 Dry and Wet Basis Calculations

	dry basis	wet basis
fraction moisture, f	$\dfrac{W_{\text{excess water}}}{W_{\text{SSD sand}}}$	$\dfrac{W_{\text{excess water}}}{W_{\text{SSD sand}} + W_{\text{excess water}}}$
weight of sand, $W_{\text{wet sand}}$	$W_{\text{SSD sand}} + W_{\text{excess water}}$ $(1+f)W_{\text{SSD sand}}$	$W_{\text{SSD sand}} + W_{\text{excess water}}$ $\left(\dfrac{1}{1-f}\right)W_{\text{SSD sand}}$
weight of SSD sand, $W_{\text{SSD sand}}$	$\dfrac{W_{\text{wet sand}}}{1+f}$	$(1-f)W_{\text{wet sand}}$
weight of excess water, $W_{\text{excess water}}$	$fW_{\text{SSD sand}}$ $\dfrac{fW_{\text{wet sand}}}{1+f}$	$fW_{\text{wet sand}}$ $\dfrac{fW_{\text{SSD sand}}}{1-f}$

Table 49.4 Values of the Unit Weight Coefficient, C_w

unit weight, γ, (density, ρ) of concrete	C_w
<140 lbf/ft³ (<2240 kg/m³)	$0.5\left(1+\dfrac{\gamma_{\text{lbf/ft}^3}}{145}\right) \geq 0.8$ $\left(0.5\left(1+\dfrac{\rho_{\text{kg/m}^3}}{2320}\right) \geq 0.8\right)$
140 lbf/ft³ $\leq \gamma \leq 150$ lbf/ft³ (2240 kg/m³ $\leq \rho \leq 2400$ kg/m³)	1.0
>150 lbf/ft³ (>2400 kg/m³)	$\dfrac{\gamma_{\text{lbf/ft}^3}}{145}$ $\left(\dfrac{\rho_{\text{kg/m}^3}}{2320}\right)$

Table 49.5 Values of the Chemistry Coefficient, C_c

cement type or blend	C_c
types I, II, and III without retarders	1.0
types I, II, and III with retarders	1.2
other types or blends containing less than 70% slag or less than 40% fly ash without retarders	1.2
other types or blends containing less than 70% slag or less than 40% fly ash with retarders	1.4
blends containing more than 70% slag or more than 40% fly ash	1.4

CERM Chapter 50
Reinforced Concrete: Beams
(Only the unified design methods presented in ACI 318 Chap. 5 may be used on the exam.)

5. SERVICE LOADS, FACTORED LOADS, AND LOAD COMBINATIONS

$$U = 1.4D \qquad 50.1$$
$$U = 1.2D + 1.6L \qquad 50.2$$
$$U = 1.2D + 0.5W \qquad 50.3$$
$$U = 1.2D + 1.0W + 1.0L \qquad 50.4$$
$$U = 0.9D + 1.0W \qquad 50.5$$
$$U = 1.2D + 1.0E + 1.0L \qquad 50.6$$
$$U = 0.9D + 1.0E \qquad 50.7$$

6. DESIGN STRENGTH AND DESIGN CRITERIA

$$\text{design strength} = \phi(\text{nominal strength}) \qquad 50.8$$
$$\phi M_n \geq M_u \qquad 50.9(a)$$
$$\phi V_n \geq V_u \qquad 50.9(b)$$

$$p_{\text{max,kPa}} = 30C_w \leq C_wC_c$$
$$\times \left(7.2 + \dfrac{1156 + 244R_{\text{m/h}}}{T_{°C} + 17.8°}\right) \leq \rho gh$$
$$\begin{bmatrix}\text{walls: } h>4.2 \text{ m, } R\leq2.1 \text{ m/h}\\ \text{walls: } 2.1 \text{ m/h} < R \leq 4.5 \text{ m/h}\end{bmatrix} \text{[SI]}$$
$$49.4(a)$$

$$p_{\text{max,psf}} = 600C_w \leq C_wC_c$$
$$\times \left(150 + \dfrac{43{,}400 + 2800R_{\text{ft/hr}}}{T_{°F}}\right) \leq \gamma h$$
$$\begin{bmatrix}\text{walls: } h>14 \text{ ft, } R\leq7 \text{ ft/hr}\\ \text{walls: } 7 \text{ ft/hr} < R \leq 15 \text{ ft/hr}\end{bmatrix} \text{[U.S.]}$$
$$49.4(b)$$

- *Tension controlled ($\epsilon_t > 0.005$):*

$$\phi = 0.90$$

- *Compression controlled ($\epsilon_t \leq 0.002$):*

$$\phi = 0.75 \text{ (spiral reinforcement)}$$

$$\phi = 0.65 \text{ (other reinforcement)}$$

- *Transition region ($0.002 < \epsilon_t \leq 0.005$):*

For spiral reinforcement,

$$\phi = 0.75 + 50(\epsilon_t - 0.002) \quad [\epsilon_t \leq 0.005]$$
$$\text{[ACI 318 Fig. R21.2.2b]} \quad \textit{50.10}$$

For other reinforcement,

$$\phi = 0.65 + \frac{250}{3}(\epsilon_t - 0.002) \quad [\epsilon_t \leq 0.005]$$
$$\text{[ACI 318 Fig. R21.2.2b]} \quad \textit{50.11}$$

- *Shear:*

$$\phi = 0.75$$

7. MINIMUM STEEL AREA

$$A_{s,\min} = \frac{\sqrt{f'_c} b_w d}{4 f_y} \geq \frac{1.4 b_w d}{f_y} \qquad \text{[SI]} \quad \textit{50.13(a)}$$

$$A_{s,\min} = \frac{3\sqrt{f'_c} b_w d}{f_y} \geq \frac{200 b_w d}{f_y} \qquad \text{[U.S.]} \quad \textit{50.13(b)}$$

$$A_{s,\min} = \frac{\sqrt{f'_c} b_w d}{2 f_y} \qquad \text{[SI]} \quad \textit{50.14(a)}$$

$$A_{s,\min} = \frac{6\sqrt{f'_c} b_w d}{f_y} \qquad \text{[U.S.]} \quad \textit{50.14(b)}$$

8. MAXIMUM STEEL AREA

$$A_{sb} = \rho_{sb} bd = \frac{0.85 f'_c A_{cb}}{f_y} = \frac{0.85 f'_c a_b b}{f_y}$$
$$= \frac{0.85 f'_c \beta_1 c_b b}{f_y} \qquad \textit{50.15}$$

$$a_b = \beta_1 \left(\frac{600}{600 + f_y}\right) d \qquad \text{[SI]} \quad \textit{50.16(a)}$$

$$a_b = \beta_1 \left(\frac{87,000}{87,000 + f_y}\right) d \qquad \text{[U.S.]} \quad \textit{50.16(b)}$$

$$\beta_1 = 0.85 \qquad [f'_c \leq 27.6 \text{ MPa}] \qquad \text{[SI]} \quad \textit{50.17(a)}$$

$$\beta_1 = 0.85 \qquad [f'_c \leq 4000 \text{ lbf/in}^2] \qquad \text{[U.S.]} \quad \textit{50.17(b)}$$

$$\beta_1 = 0.85 - 0.05\left(\frac{f'_c - 27.6}{6.9}\right) \geq 0.65 \qquad \text{[SI]} \quad \textit{50.17(c)}$$

$$\beta_1 = 0.85 - 0.05\left(\frac{f'_c - 4000}{1000}\right) \geq 0.65 \qquad \text{[U.S.]} \quad \textit{50.17(d)}$$

$$A_{sb} = \rho_{sb} bd = \left(\frac{0.85 \beta_1 f'_c}{f_y}\right)\left(\frac{600}{600 + f_y}\right) bd$$
$$\text{[SI]} \quad \textit{50.18(a)}$$

$$A_{sb} = \rho_{sb} bd = \left(\frac{0.85 \beta_1 f'_c}{f_y}\right)\left(\frac{87,000}{87,000 + f_y}\right) bd$$
$$\text{[U.S.]} \quad \textit{50.18(b)}$$

$$A_{s,\max} = 0.75 A_{sb} \qquad \text{[alternative provisions only]} \qquad \textit{50.19}$$

Figure 50.5 *Beam at Balanced Condition*

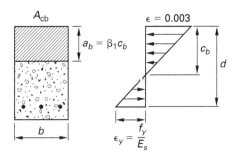

9. STEEL COVER AND BEAM WIDTH

$$1.5 \leq d/b \leq 2.5 \qquad \textit{50.20}$$

Table 50.2 *Minimum Beam Widths[a,b] for Beams with $1^1/2$ in Cover (inches)*

size of bar	number of bars in a single layer of reinforcement							add for each additional bar[c]
	2	3	4	5	6	7	8	
no. 4	6.8	8.3	9.8	11.3	12.8	14.3	15.8	1.50
no. 5	6.9	8.5	10.2	11.8	13.4	15.0	16.7	1.63
no. 6	7.0	8.8	10.5	12.3	14.0	15.8	17.5	1.75
no. 7	7.2	9.0	10.9	12.8	14.7	16.5	18.4	1.88
no. 8	7.3	9.3	11.3	13.3	15.3	17.3	19.3	2.00
no. 9	7.6	9.8	12.2	14.3	16.6	18.8	21.1	2.26
no. 10	7.8	10.4	12.9	15.5	18.0	20.5	23.1	2.54
no. 11	8.1	10.9	13.8	16.6	19.4	22.2	25.0	2.82
no. 14	8.9	12.3	15.7	19.1	22.5	25.9	29.3	3.40
no. 18	10.6	15.1	19.6	24.1	28.6	33.1	37.6	4.51

(Multiply in by 25.4 to obtain mm.)
[a]Using no. 3 stirrups. If stirrups are not used, deduct 0.75 in.
[b]The minimum inside radius of a 90° stirrup bend is two times the stirrup diameter. An allowance has been included in the beam widths to achieve a full bend radius.
[c]For additional horizontal bars, the beam width is increased by adding the value in the last column.

10. NOMINAL MOMENT STRENGTH OF SINGLY REINFORCED SECTIONS

$$M_n = f_y A_s (\text{lever arm}) \qquad 50.21$$

$$\epsilon_t = 0.003 \left(\frac{d-c}{c} \right) \qquad 50.22$$

$$f_s = (600 \text{ MPa}) \left(\frac{d-c}{c} \right) \qquad \text{[SI]} \quad 50.23(a)$$

$$f_s = \left(87,000 \, \frac{\text{lbf}}{\text{in}^2} \right) \left(\frac{d-c}{c} \right) \qquad \text{[U.S.]} \quad 50.23(b)$$

$$\rho = \frac{A_s}{bd} = 0.85 \beta_1 \left(\frac{c}{d} \right) \left(\frac{f_c'}{f_y} \right) \qquad 50.24$$

$$T = f_y A_s \qquad 50.25$$

$$A_c = \frac{T}{0.85 f_c'} = \frac{f_y A_s}{0.85 f_c'} \qquad 50.26$$

$$M_n = A_s f_y (d - \lambda) \qquad 50.27$$

Figure 50.6 *Conditions at Maximum Moment*

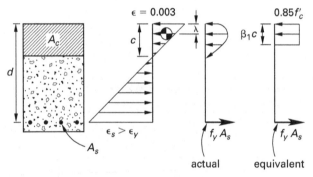

(a) strain distribution (b) compressive stress distribution (c) equivalent rectangular compressive stress block

11. BEAM DESIGN: SIZE KNOWN, REINFORCEMENT UNKNOWN

$$A_s = \frac{M_u}{\phi f_y (d - \lambda)} \qquad 50.28$$

12. BEAM DESIGN: SIZE AND REINFORCEMENT UNKNOWN

(See Table 50.3.)

$$\lambda = \frac{A_c}{2b} \qquad 50.29$$

$$M_n = A_s f_y d \left(1 - \frac{A_s f_y}{1.7 f_c' b d} \right) \qquad 50.30$$

$$\rho = \frac{A_s}{bd} \qquad 50.31$$

$$M_n = \rho b d^2 f_y \left(1 - \frac{\rho f_y}{1.7 f_c'} \right)$$
$$= A_s f_y \left(d - \frac{A_s f_y}{1.7 f_c' b} \right) \qquad 50.32$$

$$\rho_{\text{sb}} = \left(\frac{0.85 \beta_1 f_c'}{f_y} \right) \left(\frac{600}{600 + f_y} \right) \qquad \text{[SI]} \quad 50.33(a)$$

$$\rho_{\text{sb}} = \left(\frac{0.85 \beta_1 f_c'}{f_y} \right) \left(\frac{87,000}{87,000 + f_y} \right) \qquad \text{[U.S.]} \quad 50.33(b)$$

$$\rho_{\text{min}} = \frac{\sqrt{f_c'}}{4 f_y} \geq \frac{1.4}{f_y} \qquad \text{[SI]} \quad 50.34(a)$$

$$\rho_{\text{min}} = \frac{3 \sqrt{f_c'}}{f_y} \geq \frac{200}{f_y} \qquad \text{[U.S.]} \quad 50.34(b)$$

13. SERVICEABILITY: CRACKING

(See Fig. 50.7.)

$$s_{\text{max}} = 15 \left(\frac{40,000}{f_s} \right) - 2.5 c_c \leq 12 \left(\frac{40,000}{f_s} \right) \qquad 50.36$$

14. CRACKED MOMENT OF INERTIA

(See Fig. 50.8.)

$$n = \frac{E_s}{E_c} \qquad 50.37$$

$$E_c = 4700 \sqrt{f_c'} \qquad \text{[SI]} \quad 50.38(a)$$

$$E_c = 57,000 \sqrt{f_c'} \qquad \text{[U.S.]} \quad 50.38(b)$$

$$\frac{b c_s^2}{2} = n A_s (d - c_s) \qquad 50.39$$

$$c_s = \frac{n A_s}{b} \left(\sqrt{1 + \frac{2bd}{n A_s}} - 1 \right) \qquad 50.40$$

$$c_s = n \rho d \left(\sqrt{1 + \frac{2}{n \rho}} - 1 \right) \qquad 50.41$$

$$I_{\text{cr}} = \frac{b c_s^3}{3} + n A_s (d - c_s)^2 \qquad 50.42$$

Table 50.3 Total Areas for Various Numbers of Bars (in²)

bar size	nominal diameter (in)	weight (lbf/ft)	number of bars									
			1	2	3	4	5	6	7	8	9	10
no. 3	0.375	0.376	0.11	0.22	0.33	0.44	0.55	0.66	0.77	0.88	0.99	1.10
no. 4	0.500	0.668	0.20	0.40	0.60	0.80	1.00	1.20	1.40	1.60	1.80	2.00
no. 5	0.625	1.043	0.31	0.62	0.93	1.24	1.55	1.86	2.17	2.48	2.79	3.10
no. 6	0.750	1.502	0.44	0.88	1.32	1.76	2.20	2.64	3.08	3.52	3.96	4.40
no. 7	0.875	2.044	0.60	1.20	1.80	2.40	3.00	3.60	4.20	4.80	5.40	6.00
no. 8	1.000	2.670	0.79	1.58	2.37	3.16	3.95	4.74	5.53	6.32	7.11	7.90
no. 9	1.128	3.400	1.00	2.00	3.00	4.00	5.00	6.00	7.00	8.00	9.00	10.0
no. 10	1.270	4.303	1.27	2.54	3.81	5.08	6.35	7.62	8.89	10.16	11.43	12.70
no. 11	1.410	5.313	1.56	3.12	4.68	6.24	7.80	9.36	10.92	12.48	14.04	15.60
no. 14[*]	1.693	7.650	2.25	4.50	6.75	9.00	11.25	13.5	15.75	18.00	20.25	22.50
no. 18[*]	2.257	13.60	4.00	8.00	12.0	16.0	20.00	24.0	28.00	32.00	36.00	40.00

(Multiply in by 25.4 to obtain mm.)
(Multiply in² by 645 to obtain mm².)
(Multiply lbf/ft by 1.488 to obtain kg/m.)
[*]Number 14 and no. 18 bars are typically used in columns only.

Figure 50.7 Parameters for Cracking Calculation

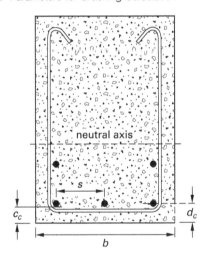

Figure 50.8 Parameters for Cracked Moment of Inertia

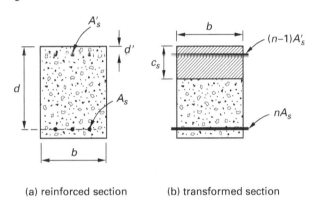

(a) reinforced section   (b) transformed section

15. SERVICEABILITY: DEFLECTIONS

$$I_e = \left(\frac{M_{cr}}{M_a}\right)^3 I_g + \left(1 - \left(\frac{M_{cr}}{M_a}\right)^3\right) I_{cr} \leq I_g \qquad 50.43$$

$$M_{cr} = \frac{f_r I_g}{y_t} \qquad 50.44$$

$$f_r = 0.7\lambda\sqrt{f_c'} \qquad \text{[SI]} \quad 50.45(a)$$

$$f_r = 7.5\lambda\sqrt{f_c'} \qquad \text{[U.S.]} \quad 50.45(b)$$

16. LONG-TERM DEFLECTIONS

$$\Delta_a = \lambda\Delta_i \qquad 50.46$$

$$\lambda = \frac{i}{1 + 50\rho'} \qquad 50.47$$

17. MINIMUM BEAM DEPTHS TO AVOID EXPLICIT DEFLECTION CALCULATIONS

The values in Table 50.4 apply to normalweight concrete reinforced with $f_y = 60{,}000$ lbf/in² (400 MPa) steel. The table values are multiplied by the adjustment factor in Eq. 50.48 for other steel strengths.

$$0.4 + \frac{f_y}{700} \qquad \text{[SI]} \quad 50.48(a)$$

$$0.4 + \frac{f_y}{100{,}000} \qquad \text{[U.S.]} \quad 50.48(b)$$

For structural lightweight concrete with a unit weight, w, between 90 lbf/ft³ and 115 lbf/ft³ (14.13 kN/m³ and 18.07 kN/m³), multiply the Table 50.4 values by

$$1.65 - 0.0318w \geq 1.09 \qquad \text{[SI]} \quad 50.49(a)$$

$$1.65 - 0.005w \geq 1.09 \qquad \text{[U.S.]} \quad 50.49(b)$$

Table 50.4 *Minimum Beam Thickness Unless Deflections Are Computed (f_y = 60,000 lbf/in²)**

construction	minimum h (fraction of span length)
simply supported	$\frac{1}{16}$
one end continuous	$\frac{1}{18.5}$
both ends continuous	$\frac{1}{21}$
cantilever	$\frac{1}{8}$

*Corrections are required for other steel strengths. See Eq. 50.48 and Eq. 50.49.

Source: ACI 318 Table 9.3.1.1

19. DESIGN OF T-BEAMS

case 1: Beams with flanges on each side of the web [ACI 318 Sec. 6.3.2.1]

Effective width (including the compression area of the stem) for a T-beam may not exceed one-eighth of the beam's span length.

case 2: Beams with an L-shaped flange [ACI 318 Sec. 6.3.2.1]

Effective overhanging flange width (excluding the compression area of the stem) is the minimum of

one-twelfth of the beam's span length, or six times the slab's thickness, or one-half of the clear distance between beam webs

case 3: Isolated T-beams [ACI 318 Sec. 6.3.2.1]

If a T-beam is not part of a floor system but the T shape is used to provide additional compression area, the flange thickness must not be less than one-half of the width of the web. The effective width of the flange must not be more than four times the width of the web.

When a T-beam is subjected to a negative moment, the tension steel must be placed in the flange. ACI 318 Sec. 24.3.4 requires that this tension steel not all be placed inside the region of the web but that it be spread out into the flange over a distance whose width is the smaller of (a) the effective flange width or (b) one-tenth of the span. This provision ensures that cracking on the top surface will be distributed.

If the T-beam is an isolated member, loading will produce bending of the flange in the direction perpendicular to the beam span. Adequate transverse steel must be provided to prevent a bending failure of the flange.

21. SHEAR REINFORCEMENT

$$\phi V_n = V_u \qquad 50.50$$
$$V_n = V_c + V_s \qquad 50.51$$

22. SHEAR STRENGTH PROVIDED BY CONCRETE

$$V_c = \left(\frac{\lambda\sqrt{f_c'}}{7} + \frac{120}{7}\rho_w\left(\frac{V_u d}{M_u}\right)\right)b_w d \le 0.29\lambda\sqrt{f_c'}b_w d$$
[SI]   50.52(a)

$$V_c = \left(1.9\lambda\sqrt{f_c'} + 2500\rho_w\left(\frac{V_u d}{M_u}\right)\right)b_w d \le 3.5\lambda\sqrt{f_c'}b_w d$$
[U.S.]   50.52(b)

$$V_c = \tfrac{1}{6}\lambda\sqrt{f_c'}b_w d \qquad [SI] \quad 50.53(a)$$
$$V_c = 2\lambda\sqrt{f_c'}b_w d \qquad [U.S.] \quad 50.53(b)$$

23. SHEAR STRENGTH PROVIDED BY SHEAR REINFORCEMENT

$$V_s = \frac{A_v f_{yt} d}{s} \qquad 50.54$$
$$V_s = \frac{A_v f_{yt}(\sin\theta + \cos\theta)d}{s} \qquad 50.55$$
$$V_s = A_v f_y \sin\theta \le 0.25\sqrt{f_c'}b_w d \qquad [SI] \quad 50.56(a)$$
$$V_s = A_v f_y \sin\theta \le 3\sqrt{f_c'}b_w d \qquad [U.S.] \quad 50.56(b)$$

24. SHEAR REINFORCEMENT LIMITATIONS

$$V_s \le \tfrac{2}{3}\sqrt{f_c'}b_w d \quad [\text{steel only}] \qquad [SI] \quad 50.57(a)$$
$$V_s \le 8\sqrt{f_c'}b_w d \quad [\text{steel only}] \qquad [U.S.] \quad 50.57(b)$$
$$V_{u,\max} = \left(\tfrac{2}{3}+\tfrac{\lambda}{6}\right)\phi\sqrt{f_c'}b_w d \quad [\text{concrete and steel}]$$
[SI]   50.58(a)
$$V_{u,\max} = (8+2\lambda)\phi\sqrt{f_c'}b_w d \quad [\text{concrete and steel}]$$
[U.S.]   50.58(b)

25. STIRRUP SPACING

$$s_{\max} = \min\{60 \text{ cm or } d/2\} \quad [V_s \le \tfrac{2}{3}\sqrt{f'_c}b_w d]$$
$$\text{[SI]} \quad 50.59(a)$$

$$s_{\max} = \min\{24 \text{ in or } d/2\} \quad [V_s \le 4\sqrt{f'_c}b_w d]$$
$$\text{[U.S.]} \quad 50.59(b)$$

$$s_{\max} = \min\{30 \text{ cm or } d/4\} \quad [V_s > \tfrac{2}{3}\sqrt{f'_c}b_w d]$$
$$\text{[SI]} \quad 50.60(a)$$

$$s_{\max} = \min\{12 \text{ in or } d/4\} \quad [V_s > 4\sqrt{f'_c}b_w d]$$
$$\text{[U.S.]} \quad 50.60(b)$$

$$A_{v,\min} = \tfrac{1}{8}\sqrt{f'_c}\left(\frac{b_w s}{f_{yt}}\right) \ge \frac{b_w s}{3f_{yt}} \quad \text{[SI]} \quad 50.61(a)$$

$$A_{v,\min} = 0.75\sqrt{f'_c}\left(\frac{b_w s}{f_{yt}}\right) \ge \frac{50b_w s}{f_{yt}} \quad \text{[U.S.]} \quad 50.61(b)$$

26. NO-STIRRUP CONDITIONS

$$\frac{\phi V_c}{2} \ge V_u \qquad 50.62$$

27. SHEAR REINFORCEMENT DESIGN PROCEDURE

$$V_{s,\text{req}} = \frac{V_u}{\phi} - V_c \qquad 50.63$$

$$s = \frac{A_v f_{yt} d}{V_{s,\text{req}}} \qquad 50.64$$

28. ANCHORAGE OF SHEAR REINFORCEMENT

$$\text{embedment length} \ge \frac{0.17 d_b f_{yt}}{\lambda\sqrt{f'_c}} \quad \text{[SI]} \quad 50.65(a)$$

$$\text{embedment length} \ge \frac{0.014 d_b f_{yt}}{\lambda\sqrt{f'_c}} \quad \text{[U.S.]} \quad 50.65(b)$$

30. STRENGTH ANALYSIS OF DOUBLY REINFORCED SECTIONS

$$A_c = \frac{f_y A_s - f'_s A'_s}{0.85 f'_c} \qquad 50.66$$

$$\epsilon'_s = \left(\frac{0.003}{c}\right)(c - d') \qquad 50.67$$

$$f'_s = E_s \epsilon'_s \le f_y \qquad 50.68$$

$$M_n = (A_s f_y - A'_s f'_s)(d - \lambda) + A'_s f'_s (d - d') \qquad 50.69$$

31. DESIGN OF DOUBLY REINFORCED SECTIONS

$$M'_n = \frac{M_u}{\phi} - M_{\text{nc}} \qquad 50.70$$

$$A_{s,\text{add}} = \frac{M'_n}{f_y(d - d')} \qquad 50.71$$

$$A_s = A_{s,\max} + A_{s,\text{add}} \qquad 50.72$$

$$A'_s = \left(\frac{f_y}{f'_s}\right)A_{s,\text{add}} \qquad 50.73$$

CERM Chapter 51
Reinforced Concrete: Slabs

3. TEMPERATURE STEEL

$$\rho_t = 0.0018\left(\frac{413.7}{f_y}\right) \quad \text{[SI]} \quad 51.1(a)$$

$$\rho_t = 0.0018\left(\frac{60,000}{f_y}\right) \quad \text{[U.S.]} \quad 51.1(b)$$

6. SLAB DESIGN FOR FLEXURE

$$s = \left(\frac{A_b}{A_{\text{sr}}}\right)\left(12 \frac{\text{in}}{\text{ft}}\right) \qquad 51.2$$

9. DIRECT DESIGN METHOD

$$M_o = \frac{q_u l_2 l_n^2}{8} \quad \text{[ACI Eq. 13-4]} \qquad 51.3$$

10. FACTORED MOMENTS IN SLAB BEAMS

(See Fig. 51.4.)

$$\alpha = \frac{E_{cb} I_b}{E_{cs} I_s} \qquad 51.4$$

11. COMPUTATION OF RELATIVE TORSIONAL STIFFNESS

$$\beta_t = \frac{E_{cb} C}{2 E_{cs} I_s} \qquad 51.5$$

$$C = \sum\left(1 - 0.63\left(\frac{x}{y}\right)\right)\left(\frac{x^3 y}{3}\right) \qquad 51.6$$

Figure 51.4 *Two-Way Slab Beams (monolithic or fully composite construction) (ACI 318 Sec. 8.4.1.8)*

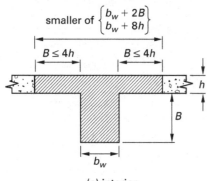

(a) interior

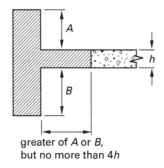

greater of *A* or *B*, but no more than 4*h*

(b) exterior

12. DEFLECTIONS IN TWO-WAY SLABS

$$h = \frac{l_n\left(0.8 + \dfrac{f_y}{200{,}000}\right)}{36 + 5\beta(\alpha_m - 0.2)}$$
$$\geq 5 \text{ in} \qquad \qquad 51.7$$

$$h = \frac{l_n\left(0.8 + \dfrac{f_y}{200{,}000}\right)}{36 + 9\beta}$$
$$\geq 3.5 \text{ in} \qquad \qquad 51.8$$

CERM Chapter 52
Reinforced Concrete: Short Columns
(Only the unified design methods presented in ACI 318 Chap. 5 may be used on the exam.)

1. INTRODUCTION

$$\frac{k_b l_u}{r} \leq 34 - 12\left(\frac{M_1}{M_2}\right) \leq 40 \qquad 52.1$$

$$\frac{k_u l_u}{r} \leq 22 \quad [k_u > 1] \qquad 52.2$$

2. TIED COLUMNS

$$0.01 \leq \rho_g \leq 0.08 \qquad 52.3$$

3. SPIRAL COLUMNS

$$\rho_s = 0.45\left(\frac{A_g}{A_c} - 1\right)\left(\frac{f'_c}{f_{yt}}\right) \qquad 52.4$$

$$s \approx \frac{4A_{sp}}{\rho_s D_c} \qquad 52.5$$

$$\text{clear distance} = s - d_{sp} \qquad 52.6$$

4. DESIGN FOR SMALL ECCENTRICITY

$$\phi\beta P_o \geq P_u \qquad 52.7$$

$$P_o = 0.85f'_c(A_g - A_{st}) + f_y A_{st} \qquad 52.8$$

$$P_u = 1.2(\text{dead load axial force})$$
$$+ 1.6(\text{live load axial force}) \qquad 52.9$$

$$0.85f'_c(A_g - A_{st}) + A_{st}f_y = \frac{P_u}{\phi\beta} \qquad 52.10$$

$$A_g\left(0.85f'_c(1 - \rho_g) + \rho_g f_y\right) = \frac{P_u}{\phi\beta} \qquad 52.11$$

$$\rho_g = \frac{A_{st}}{A_g} \qquad 52.12$$

CERM Chapter 53
Reinforced Concrete: Long Columns

2. BRACED AND UNBRACED COLUMNS

$$Q = \frac{\sum P_u \Delta_o}{V_{us} l_c} \leq 0.05 \qquad 53.1$$

3. EFFECTIVE LENGTH

$$\text{effective length} = kl \qquad 53.2$$

$$\Psi = \frac{\displaystyle\sum_{\text{columns}} \frac{EI}{l_c}}{\displaystyle\sum_{\text{beams}} \frac{EI}{l}} \qquad 53.3$$

4. FRAME SECTION PROPERTIES (SECOND-ORDER ANALYSIS)

For compression members,

$$I = \left(0.80 + 25\frac{A_{st}}{A_g}\right)\left(1 - \frac{M_u}{P_u h} - 0.5\frac{P_u}{P_o}\right)I_g \qquad 53.4$$

For flexural members,

$$I = (0.10 + 25\rho)\left(1.2 - 0.2\frac{b_w}{d}\right)I_g \qquad 53.5$$

5. EFFECTIVE FLEXURAL STIFFNESS: COLUMNS IN BRACED FRAMES

$$EI = \frac{0.2E_cI_g + E_sI_{se}}{1 + \beta_{dns}} \quad \text{[braced frames]} \qquad 53.6$$

$$EI = \frac{0.4E_cI_g}{1 + \beta_{dns}} \quad \text{[braced frames]} \qquad 53.7$$

6. BUCKLING LOAD

$$P_c = \frac{\pi^2 EI}{(kl_u)^2} \quad \text{[ACI 318 Eq. 6.6.4.4.2]} \qquad 53.8$$

7. COLUMNS IN BRACED STRUCTURES (NON-SWAY FRAMES)

$$M_c = \delta M_2 \qquad 53.9$$

$$\delta = \frac{C_m}{1 - \dfrac{P_u}{0.75P_c}} \geq 1.0 \qquad 53.10$$

$$C_m = 0.6 + 0.4\left(\frac{M_1}{M_2}\right) \qquad 53.11$$

$$M_{2,min,in-lbf} = P_{u,lbf}(0.6 + 0.03h_{in}) \qquad 53.12$$

8. COLUMNS IN UNBRACED STRUCTURES (SWAY FRAMES)

$$M_1 = M_{1ns} + \delta_s M_{1s} \quad \text{[ACI 318 Eq. 6.6.4.6.1a]} \qquad 53.13$$

$$M_2 = M_{2ns} + \delta_s M_{2s} \quad \text{[ACI 318 Eq. 6.6.4.6.1b]} \qquad 53.14$$

$$\delta_s = \frac{1}{1-Q} \geq 1 \quad \text{[ACI 318 Eq. 6.6.4.6.2a]} \qquad 53.15$$

$$\delta_s = \frac{1}{1 - \dfrac{\sum P_u}{0.75\sum P_c}} \geq 1.0 \quad \text{[ACI 318 Eq. 6.6.4.6.2b]}$$

$$53.16$$

CERM Chapter 54
Reinforced Concrete: Walls and Retaining Walls

(Only the unified design methods presented in ACI 318 Chap. 5 may be used on the exam.)

2. BEARING WALLS: EMPIRICAL METHOD

$$P_u \leq \phi P_{n,w} \leq 0.55\phi f'_c A_g \left(1 - \left(\frac{kl_c}{32h}\right)^2\right) \qquad 54.1$$

5. DESIGN OF RETAINING WALLS

$$\rho \approx 0.02 \qquad 54.2$$

$$R_{u,trial} = \rho f_y \left(1 - \left(\frac{\rho f_y}{(2)(0.85)f'_c}\right)\right) \qquad 54.3$$

$$M_{u,stem} = 1.6R_{a,h}y_a \qquad 54.4$$

$$d = \sqrt{\frac{M_{u,stem}}{\phi R_u w}} \qquad 54.5$$

$$t_{stem} = d + \text{cover} + \tfrac{1}{2}d_b \qquad 54.6$$

$$V_{u,stem} = 1.6V_{active} + 1.6V_{surcharge} \qquad 54.7$$

$$\phi V_n = 2\phi w d' \lambda \sqrt{f'_c} \qquad 54.8$$

$$V_{u,base} = 1.2V_{soil} + 1.2V_{heel weight} + 1.6V_{surcharge} \qquad 54.9$$

Figure 54.1 Retaining Wall Dimensions (step 1)

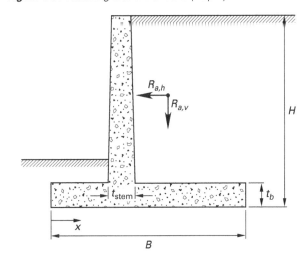

Figure 54.2 Base of Stem Details (step 5)

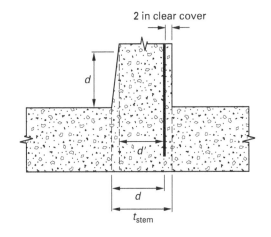

$$M_{u,\text{base}} = 1.2M_{\text{soil}} + 1.2M_{\text{heel weight}}$$
$$+ 1.6M_{\text{surcharge}} \qquad 54.10$$

$$V_{u,\text{toe}} = 1.6V_{\text{toe pressure}} - 1.2V_{\text{toe weight}} \qquad 54.11$$

$$M_{u,\text{toe}} = 1.6M_{\text{toe pressure}} - 1.2M_{\text{toe weight}} \qquad 54.12$$

$$M_n = \frac{M_u}{\phi} = \rho f_y \left(1 - \frac{\rho f_y}{(2)(0.85)f_c'}\right)bd^2 \qquad 54.13$$

Figure 54.3 *Heel Details (step 9)*

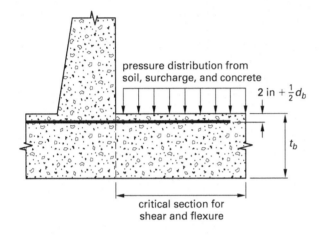

pressure distribution from
soil, surcharge, and concrete

2 in + $\frac{1}{2}d_b$

t_b

critical section for
shear and flexure

CERM Chapter 55
Reinforced Concrete: Footings
(Only the unified design methods presented in
ACI 318 Chap. 5 may be used on the exam.)

1. INTRODUCTION

$$q_s = \frac{P_s}{A_f} + \gamma_c h + \gamma_s(H - h) \pm \frac{M_s\left(\frac{B}{2}\right)}{I_f} \le q_a \qquad 55.1$$

$$I_f = \tfrac{1}{12}LB^3 \qquad 55.2$$

2. WALL FOOTINGS

$$q_u = \frac{P_u}{B} \qquad 55.3$$

$$P_u = 1.2P_d + 1.6P_l \qquad 55.4$$

$$\phi V_n \ge V_u \qquad 55.5$$

$$V_n = V_c + V_s \qquad 55.6$$

$$V_c = v_c L d = 2\lambda\sqrt{f_c'}Ld \qquad 55.7$$

$$v_u = \frac{q_u}{d}\left(\frac{B-t}{2} - d\right) \qquad 55.8$$

$$d = \frac{q_u(B-t)}{2(q_u + \phi v_n)} \qquad 55.9$$

$$h = d + \tfrac{1}{2}(\text{diameter of } x \text{ bars})$$
$$+ \text{diameter of } z \text{ bars} + \text{cover}$$
$$\begin{bmatrix} \text{orthogonal reinforcement} \\ \text{under main steel} \end{bmatrix} \qquad 55.10(a)$$

$$h = d + \tfrac{1}{2}(\text{diameter of } x \text{ bars}) + \text{cover}$$
$$\begin{bmatrix} \text{orthogonal reinforcement} \\ \text{over main steel} \end{bmatrix} \qquad 55.10(b)$$

$$h \ge 6 \text{ in} + \text{diameter of } x \text{ bars}$$
$$+ \text{diameter of } z \text{ bars} + 3 \text{ in} \qquad 55.11$$

3. COLUMN FOOTINGS

$$v_u = \frac{q_u e}{d} \qquad 55.12$$

$$A_p = 2(b_1 + b_2)d \qquad 55.13$$

$$v_u = \frac{V_u}{A_c} + \frac{\gamma_v M_u c}{J_c}$$
$$= \frac{P_u - R}{A_p} + \frac{\gamma_v M_u(0.5b_1)}{J_c} \qquad 55.14$$

$$R = \frac{P_u b_1 b_2}{A_f} \qquad 55.15$$

$$\gamma_v = 1 - \frac{1}{1 + \tfrac{2}{3}\sqrt{\dfrac{b_1}{b_2}}} \qquad 55.16$$

$$J_c = \frac{db_1^3}{6}\left(1 + \left(\frac{d}{b_1}\right)^2 + 3\left(\frac{b_2}{b_1}\right)\right) \qquad 55.17$$

$$v_n = (2 + y)\lambda\sqrt{f_c'} \qquad 55.18$$

$$y = \min\{2, 4/\beta_c, 40d/b_o\} \qquad 55.19$$

Figure 55.3 *Critical Section for One-Way Shear*

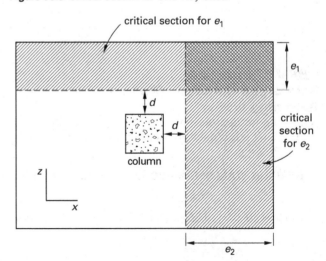

critical section for e_1

e_1

d

d

critical
section
for e_2

column

z

x

e_2

$$\beta_c = \frac{\text{column long side}}{\text{column short side}} \quad 55.20$$

$$b_o = \frac{A_p}{d} = 2(b_1 + b_2) \quad 55.21$$

$$h = d + \tfrac{1}{2}(\text{diameter of } x \text{ bars}$$
$$+ \text{diameter of } z \text{ bars}) + \text{cover} \quad 55.22$$

$$h \geq 6 \text{ in} + \text{diameter of } x \text{ bars}$$
$$+ \text{diameter of } z \text{ bars} + 3 \text{ in} \quad 55.23$$

Figure 55.4 *Critical Section for Two-Way Shear*

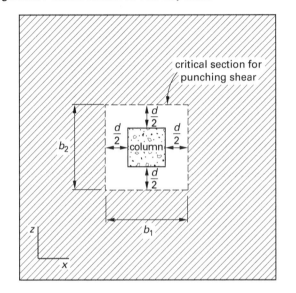

4. SELECTION OF FLEXURAL REINFORCEMENT

$$M_u = \frac{q_u L l^2}{2} \quad [\text{no column moment}] \quad 55.24$$

$$A_s = \frac{M_u}{\phi f_y(d - \lambda)} \quad 55.25$$

$$A_1 = A_{sd}\left(\frac{2}{\beta + 1}\right) \quad 55.26$$

$$A_2 = A_{sd} - A_1 \quad 55.27$$

5. DEVELOPMENT LENGTH OF FLEXURAL REINFORCEMENT

If spacing and cover satisfy the requirements given in ACI 318 Sec. 25.4.2.2,

$$l_d = \frac{d_b f_y \psi_t \psi_e}{25\lambda\sqrt{f'_c}} \geq 12 \text{ in} \quad [\text{no. 6 bars and smaller}] \quad 55.28$$

$$l_d = \frac{d_b f_y \psi_t \psi_e}{20\lambda\sqrt{f'_c}} \geq 12 \text{ in} \quad [\text{no. 7 bars and larger}] \quad 55.29$$

For all other cases,

$$l_d = \frac{3 d_b f_y \psi_t \psi_e}{50\lambda\sqrt{f'_c}} \geq 12 \text{ in} \quad \begin{bmatrix} \text{no. 6 bars and smaller;} \\ \text{Eq. 55.27 requirements} \\ \text{not met} \end{bmatrix} \quad 55.30$$

$$l_d = \frac{3 d_b f_y \psi_t \psi_e}{40\lambda\sqrt{f'_c}} \geq 12 \text{ in} \quad \begin{bmatrix} \text{no. 7 bars and larger;} \\ \text{Eq. 55.28 requirements} \\ \text{not met} \end{bmatrix} \quad 55.31$$

Using ACI 318 Sec. 25.4.2.3,

$$l_d = \frac{3 d_b f_y \psi_t \psi_e \psi_s}{40\lambda\sqrt{f'_c}\left(\dfrac{c_b + K_{tr}}{d_b}\right)} \geq 12 \text{ in} \quad 55.32$$

$$K_{tr} = \frac{40 A_{tr}}{sN} \quad 55.33$$

For hooked bars in tension,

$$l_{dh} = \frac{0.02 \psi_e d_b f_y}{\lambda\sqrt{f'_c}} \geq 8 d_b \geq 6 \text{ in} \quad \begin{bmatrix} \text{hooked} \\ \text{bars} \end{bmatrix} \quad 55.34$$

For headed deformed bars in tension,

$$l_{dt} = \frac{0.016 d_b f_y \psi_e}{\sqrt{f'_c}} \geq 8 d_b \text{ or } 6 \text{ in} \quad 55.35$$

6. TRANSFER OF FORCE AT COLUMN BASE

$$A_{db,min} = 0.005 A_g \quad 55.36$$

$$l_{dc} = \left(\frac{0.02 d_b}{\lambda}\right)\left(\frac{f_y}{\sqrt{f'_c}}\right) \geq 0.0003 d_b f_y$$
$$\geq 8 \text{ in} \quad 55.37$$

$$l_d \leq h - \text{cover} - \text{diameter of } x \text{ steel}$$
$$- \text{diameter of } z \text{ steel} \quad 55.38$$

$$P_{bearing,column} = 0.85\phi f'_{c,column} A_c \quad 55.39$$

$$P_{bearing,footing} = 0.85\alpha\phi f'_{c,footing} A_c \quad 55.40$$

$$\alpha = \sqrt{\frac{A_{ff}}{A_c}} \leq 2 \quad 55.41$$

$$A_{db} = \frac{P_u - \text{smaller of}\begin{Bmatrix} P_{bearing,column} \\ P_{bearing,footing} \end{Bmatrix}}{\phi f_y}$$
$$\geq A_{db,min} \quad 55.42$$

CERM Chapter 56
Prestressed Concrete

7. CREEP AND SHRINKAGE

$$\Delta_{\text{long term}} = \lambda\Delta_{\text{immediate}} \quad 56.1$$

$$\lambda = \frac{\xi}{1 + 50\rho'} \quad 56.2$$

8. PRESTRESS LOSSES

$$n = \frac{E_s}{E_c} \qquad 56.3$$

9. ANALYTICAL ESTIMATION OF PRESTRESS LOSSES

$$\Delta f_{\text{pLT,ksi}} = 10 \left(\frac{f_{\text{pi,ksi}} A_{\text{ps}}}{A_g} \right) \gamma_h \gamma_{\text{st}} + 12 \gamma_h \gamma_{\text{st}} + \Delta f_{\text{pR}}$$

[time-dependent losses]

56.4

$$\gamma_h = 1.7 - 0.01H \qquad 56.5$$

$$\gamma_{\text{st}} = \frac{5}{1 + f'_{\text{ci}}} \qquad 56.6$$

10. DEFLECTIONS

Figure 56.2 *Midspan Deflections from Prestressing**

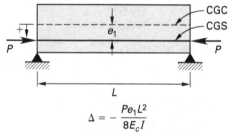

$$\Delta = -\frac{Pe_1 L^2}{8E_c I}$$

(a) straight tendons

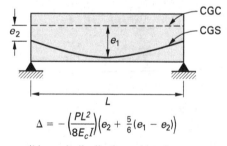

$$\Delta = -\left(\frac{PL^2}{8E_c I} \right) \left(e_2 + \tfrac{5}{6}(e_1 - e_2) \right)$$

(b) parabolically draped tendons

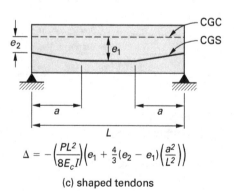

$$\Delta = -\left(\frac{PL^2}{8E_c I} \right) \left(e_1 + \tfrac{4}{3}(e_2 - e_1)\left(\tfrac{a^2}{L^2}\right) \right)$$

(c) shaped tendons

*CGC—centroid of concrete; CGS—centroid of prestressing tendons

13. MAXIMUM STRESSES

- Class U: $f_t \leq 7.5\sqrt{f'_c}$
- $f_t \leq 6\sqrt{f'_c}$ [slab systems]
- Class T: $7.5\sqrt{f'_c} < f_t \leq 12\sqrt{f'_c}$
- Class C: $f_t > 12\sqrt{f'_c}$

14. ACI PROVISIONS FOR STRENGTH

$$f_{\text{ps}} = f_{\text{pu}} \left(1 - \left(\frac{\gamma_p}{\beta_1} \right) \left(\rho_p \left(\frac{f_{\text{pu}}}{f'_c} \right) + \left(\frac{d}{d_p} \right)(\omega - \omega') \right) \right)$$

[bonded tendons]

56.8

$$\rho_p = \frac{A_{\text{ps}}}{bd_p} \qquad 56.9$$

$$\rho_p \left(\frac{f_{\text{pu}}}{f'_c} \right) + \left(\frac{d}{d_p} \right)(\omega - \omega') \geq 0.17 \qquad 56.10$$

15. ANALYSIS OF PRESTRESSED BEAMS

$$\begin{aligned} f_c &= \frac{-P}{A} - \frac{Pey}{I} + \frac{M_s y}{I} \\ &= \frac{-P}{A} - \frac{Pe}{S} + \frac{M_s}{S} \quad \text{[maximum]} \end{aligned} \qquad 56.11$$

16. SHEAR IN PRESTRESSED SECTIONS

$$V_u < \phi V_n \qquad 56.12$$

$$V_n = V_c + V_s \qquad 56.13$$

CERM Chapter 57
Composite Concrete and Steel Bridge Girders

5. EFFECTIVE SLAB WIDTH

- one-eighth of the beam span, center-to-center of supports
- one-half the distance to the centerline of the adjacent beam
- the distance to the edge of the slab

6. SECTION PROPERTIES

$$n = \frac{E_s}{E_c} \qquad 57.1$$

CERM Chapter 58
Structural Steel: Introduction

3. STEEL PROPERTIES

Table 58.1 *Typical Properties of Structural Steels*

	A992[a]/A572, grade 50	A36
modulus of elasticity, E	29,000 ksi[b]	
tensile yield strength, F_y	50 ksi	36 ksi (up to 8 in thickness)
tensile strength, F_u	65 ksi (min)	58 ksi (min)
endurance strength	30 ksi (approx)	
density, ρ	490 lbf/ft^3	
Poisson's ratio, ν	0.30 (ave)	
shear modulus, G	11,200 ksi[b]	
coefficient of thermal expansion, α	6.5×10^{-6} 1/°F (ave)	
specific heat (32–212°F)	0.107 Btu/lbm-°F	

(Multiply ksi by 6.9 to obtain MPa.)
(Multiply in by 25.4 to obtain mm.)
(Multiply lbm/ft^3 by 16 to obtain kg/m^3.)
(Multiply °F^{-1} by 9/5 to obtain °C^{-1}.)
[a]A992 steel is the *de facto* material for rolled W-shapes, having replaced A36 and A572 in most designs for new structures.
[b]as designated by AISC

CERM Chapter 59
Structural Steel: Beams

3. FLEXURAL STRENGTH IN STEEL BEAMS

$$M_n = F_y Z_x \quad \text{[AISC Eq. F2-1]} \qquad 59.1$$

$$M = M_D + M_L \le \frac{M_n}{\Omega_b} \quad \text{[ASD]} \qquad 59.2$$

$$M_u = 1.6M_L + 1.2M_D \le \phi_b M_n \quad \text{[LRFD]} \qquad 59.3$$

4. COMPACT SECTIONS

$$\frac{b_f}{2t_f} \le 0.38\sqrt{\frac{E}{F_y}} \quad \begin{bmatrix} \text{flanges in flexural} \\ \text{compression only} \end{bmatrix} \qquad 59.4$$

$$\frac{h}{t_w} \le 3.76\sqrt{\frac{E}{F_y}} \quad \begin{bmatrix} \text{webs in flexural} \\ \text{compression only} \end{bmatrix} \qquad 59.5$$

5. LATERAL BRACING

$$L_p = 1.76 r_y \sqrt{\frac{E}{F_y}} \quad \text{[AISC Eq. F2-5]} \qquad 59.6$$

$$L_r = 1.95 r_{ts} \left(\frac{E}{0.7F_y}\right)\sqrt{\frac{Jc}{S_x h_o}}$$
$$\times \sqrt{1 + \sqrt{1 + 6.76\left(\frac{0.7F_y S_x h_o}{EJc}\right)^2}}$$
$$\text{[AISC Eq. F2-6]} \qquad 59.7$$

$$r_{ts} = \frac{b_f}{\sqrt{12\left(1 + \frac{ht_w}{6b_f t_f}\right)}} \qquad 59.8$$

$$r_{ts}^2 = \frac{\sqrt{I_y C_w}}{S_x} \qquad 59.9$$

$$C_w = \frac{h^2 I_y}{4} \approx \frac{(d - t_f)^2 t_f b_f^3}{24} \qquad 59.10$$

$$J = \frac{1}{3}\int_0^{b_f} t^3 \, db \approx \frac{1}{3}\sum bt^3 \quad [b > t;\ b/t > 10] \qquad 59.11$$

For doubly symmetrical I shapes, $c = 1$. For a channel,

$$c = \frac{h_o}{2}\sqrt{\frac{I_y}{C_w}} \qquad 59.12$$

Figure 59.5 *Available Moment Versus Unbraced Length*

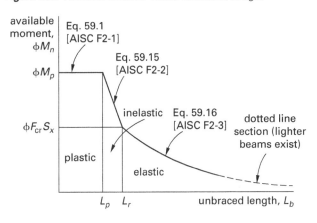

7. LATERAL TORSIONAL BUCKLING

$$M_{n,\text{nonuniform moment}} = C_b M_{n,\text{uniform moment}} \qquad 59.13$$

$$C_b = \frac{12.5 M_{\max}}{2.5 M_{\max} + 3M_A + 4M_B + 3M_C} \qquad 59.14$$

8. FLEXURAL DESIGN STRENGTH: I-SHAPES BENDING ABOUT MAJOR AXIS

If $L_r \ge L_b > L_p$, then

$$M_n = C_b\left(M_p - \left(M_p - 0.7F_y S_x\right)\left(\frac{L_b - L_p}{L_r - L_p}\right)\right)$$
$$\le M_p \quad \text{[AISC Eq. F2-2]} \qquad 59.15$$

If $L_b > L_r$, or if there is no bracing at all between support points, then

$$M_n = F_{cr}S_x \leq M_p \quad \text{[AISC Eq. F2-3]} \qquad 59.16$$

$$F_{cr} = \frac{C_b\pi^2 E}{\left(\frac{L_b}{r_{ts}}\right)^2}\sqrt{1 + 0.078\frac{Jc}{S_x h_o}\left(\frac{L_b}{r_{ts}}\right)^2}$$

$$\text{[AISC Eq. F2-4]} \qquad 59.17$$

10. SHEAR STRENGTH IN STEEL BEAMS

$$V_n = 0.6F_y A_w C_v = 0.6F_y d t_w C_v \qquad 59.18$$

$$V_u = 1.6V_L + 1.2V_D \leq \phi_v V_n \quad \text{[LRFD]} \qquad 59.19$$

$$V = V_L + V_D \leq \frac{V_n}{\Omega} \quad \text{[ASD]} \qquad 59.20$$

13. DESIGN OF STEEL BEAMS

(See Fig. 59.7.)

14. LRFD DESIGN OF CONTINUOUS BEAMS

(See Fig. 59.8.)

$$\text{factored load} = 1.2 \times \text{dead load}$$
$$+ 1.6 \times \text{live load} \qquad 59.23$$

$$Z_x = \frac{M_p}{F_y} \qquad 59.24$$

$$V_n = 0.6F_y A_w C_v \qquad 59.25$$

$$L_{pd} = \left(0.17 - 0.10\left(\frac{M_1'}{M_2}\right)\right)\left(\frac{E}{F_y}\right)r_y \geq 0.10\left(\frac{E}{F_y}\right)r_y$$
$$59.26$$

15. ULTIMATE PLASTIC MOMENTS

$$M_1 = \left(\tfrac{3}{2} - \sqrt{2}\right)wL_1^2$$
$$\approx 0.0858wL_1^2 \quad \text{[uniform loading]} \qquad 59.27$$

$$M_2 = \frac{wL^2}{16} \quad \text{[uniform loading]} \qquad 59.28$$

16. ULTIMATE SHEARS

$$V_{max} = \frac{wL}{2} + \frac{M_{max}}{L} = 0.5858wL \quad \text{[uniform loading]}$$
$$59.29$$

18. UNSYMMETRICAL BENDING

$$\left|\frac{f_a}{F_a} + \frac{f_{bx}}{F_{bx}} + \frac{f_{by}}{F_{by}}\right| \leq 1.0 \qquad 59.30$$

$$\frac{f_{bx}}{F_{bx}} + \frac{f_{by}}{F_{by}} \leq 1.0 \qquad 59.31$$

$$F_b = \frac{(\phi M)_{tabulated}}{S} \qquad 59.32$$

21. CONCENTRATED WEB FORCES

(See Fig. 59.11.)

$$R_n = F_{yw}t_w(5k + l_b) \qquad 59.33$$

$$R_n = F_{yw}t_w(2.5k + l_b) \qquad 59.34$$

For interior loads,

$$R_n = 0.80t_w^2\left(1 + 3\left(\frac{l_b}{d}\right)\left(\frac{t_w}{t_f}\right)^{1.5}\right)\sqrt{\frac{EF_{yw}t_f}{t_w}}$$
$$\text{[AISC Eq. J10-4]} \qquad 59.35$$

For $l_b/d \leq 0.2$,

$$R_n = 0.40t_w^2\left(1 + 3\left(\frac{l_b}{d}\right)\left(\frac{t_w}{t_f}\right)^{1.5}\right)\sqrt{\frac{EF_{yw}t_f}{t_w}}$$
$$\text{[AISC Eq. J10-5a]} \qquad 59.36$$

For $l_b/d > 0.2$,

$$R_n = 0.40t_w^2\left(1 + \left(\frac{4l_b}{d} - 0.2\right)\left(\frac{t_w}{t_f}\right)^{1.5}\right)\sqrt{\frac{EF_{yw}t_f}{t_w}}$$
$$\text{[AISC Eq. J10-5b]}$$
$$59.37$$

22. BEAM BEARING PLATES

(See Fig. 59.14.)

$$P_p = 0.85f_c'A_1 \quad \text{[AISC Eq. J8-1]} \qquad 59.38$$

$$A_{1,required} = \frac{P_p}{0.85\phi_p f_c'} \qquad \text{[LRFD]} \quad 59.39(a)$$

$$A_{1,required} = \frac{\Omega_p P_p}{0.85f_c'} \qquad \text{[ASD]} \quad 59.39(b)$$

Figure 59.7 *Flowchart for LRFD Beam Design (W- and M-shapes—moment, shear, and deflection criteria)*
(Spec. = Specifications; Part 16 of the AISC Manual)

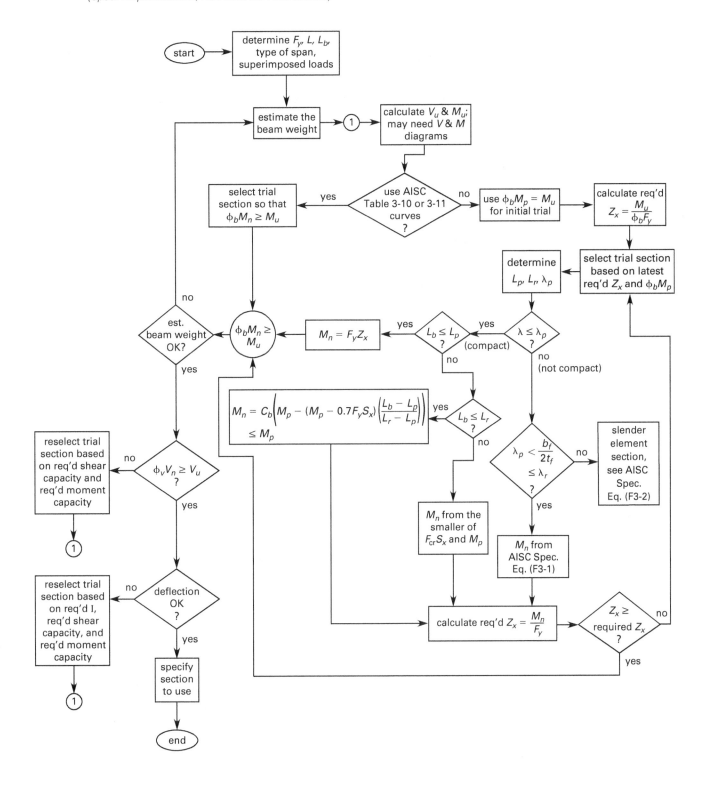

Figure 59.8 *Location of Plastic Moments on a Uniformly Loaded Continuous Beam*

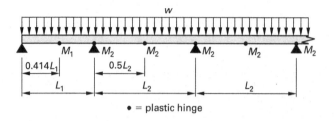

• = plastic hinge

Figure 59.11 *Nomenclature for Web Yielding Calculations*

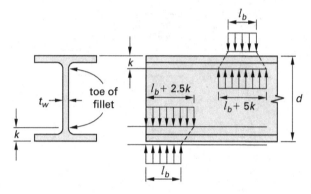

$$P_p = 0.85 f'_c A_1 \sqrt{\frac{A_2}{A_1}} \le 1.7 f'_c A_1 \quad \text{[AISC Eq. J8-2]} \quad 59.40$$

$$l_b = \frac{\phi R_n - \phi R_1}{\phi R_2} \qquad 59.41$$

$$l_b = \frac{\phi R_n - \phi R_3}{\phi R_4} \qquad 59.42$$

$$B = \frac{A_1}{l_b} \qquad 59.43$$

$$f_p = \frac{R_a}{A_1} = \frac{R_a}{B l_b} \qquad \text{[ASD]} \quad 59.44(a)$$

$$f_p = \frac{R_u}{A_1} = \frac{R_u}{B l_b} \qquad \text{[LRFD]} \quad 59.44(b)$$

Figure 59.14 *Nomenclature for Beam Bearing Plate*

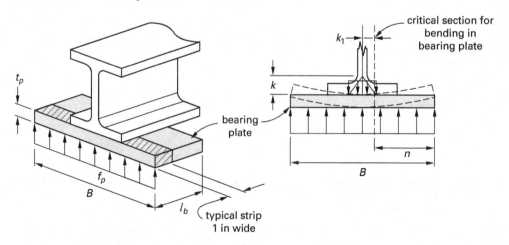

$$n = \frac{B}{2} - k_1 \qquad 59.45$$

$$t = \sqrt{\frac{2 R_a n^2 \Omega}{A_1 F_y}} = \sqrt{\frac{2 f_p n^2 \Omega}{F_y}} \qquad \text{[ASD]} \quad 59.46(a)$$

$$t = \sqrt{\frac{2 R_u n^2}{\phi A_1 F_y}} = \sqrt{\frac{2 f_p n^2}{\phi F_y}} \qquad \text{[LRFD]} \quad 59.46(b)$$

CERM Chapter 60
Structural Steel: Tension Members

2. AXIAL TENSILE STRENGTH

$$P_a = \phi_t P_n \ge P_u \qquad \text{[LRFD; design strength]} \quad 60.1(a)$$

$$P_a = \frac{P_n}{\Omega_t} \ge P \qquad \text{[ASD; allowable strength]} \quad 60.1(b)$$

$$P_n = F_y A_g \quad \begin{bmatrix} \text{yielding criterion;} \\ \textit{AISC Specification} \text{ Eq. D2-1} \end{bmatrix} \quad 60.2$$

$$P_n = F_u A_e = A_n U F_u \quad \begin{bmatrix} \text{fracture criterion;} \\ \textit{AISC Specification} \text{ Eq. D2-2} \\ \text{and Eq. D3-1} \end{bmatrix} \quad 60.3$$

3. GROSS AREA

$$A_g = bt \qquad 60.4$$

4. NET AREA

$$b_n = b - \sum d_h + \sum \frac{s^2}{4g} \qquad 60.5$$
$$A_n = b_n t \qquad 60.6$$

Figure 60.3 *Tension Member with Uniform Thickness and Unstaggered Holes*

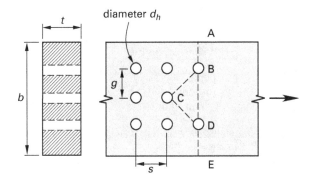

$$A_n = A_g - \sum d_h t + \left(\sum \frac{s^2}{4g} \right) t \qquad 60.7$$

5. EFFECTIVE NET AREA

(See Table 60.1 and Table 60.2.)

$$A_e = A_n U \qquad 60.8$$

$$A_n = \sum_{\substack{\text{connected} \\ \text{elements}}} A_g \qquad 60.9$$

6. BLOCK SHEAR STRENGTH

$$R_n = 0.6 F_u A_{nv} + U_{bs} F_u A_{nt}$$

$$\leq 0.6 F_y A_{gv} + U_{bs} F_u A_{nt}$$

$$\left[\begin{array}{l} \text{block shear criterion;} \\ \textit{AISC Specification} \text{ Eq. J4-5} \end{array} \right] \qquad 60.10$$

7. SLENDERNESS RATIO

$$\mathrm{SR} = \frac{L}{r_y} \qquad 60.11$$

CERM Chapter 61
Structural Steel: Compression Members

2. EULER'S COLUMN BUCKLING THEORY

$$P_e = \frac{\pi^2 EI}{L^2} \qquad 61.1$$

$$F_e = \frac{P_e}{A} = \frac{\pi^2 E}{\left(\frac{L}{r}\right)^2} \qquad 61.2$$

Table 60.1 *Shear Lag Reduction Coefficient (U) Values for Bolted Connections*[a]

shape	condition	U
all tension members where load is transmitted across each cross-sectional element by fasteners	all	1.00
W, M, S, and HP, and structural tees cut from them[b]	$b_f \geq \frac{2}{3}d$, only flange connections, and 3 or more fasteners per line in direction of loading	0.90
	$b_f < \frac{2}{3}d$, only flange connections, and 3 or more fasteners per line in direction of loading	0.85
	only web connections, and 4 or more fasteners per line in direction of loading	0.70
single angles[b]	4 or more fasteners per line in direction of loading	0.80
	2 or 3 fasteners per line in direction of loading	0.60

[a]When the connection contains only a single fastener or a single row of fasteners (i.e., $s = 0$), *AISC Commentary* Sec. J4.3 implies $U = 1$ when the tension is uniform across the face, and $U = 0.5$ otherwise; and block shear and bearing may control the connection.
[b]U may also be calculated as $U = 1 - (\overline{x}/l)$, and the larger of the two values may be used. $\overline{x}$ is the connection eccentricity, equal to the distance from the plane of the connection to the centroid of the connected member, often tabulated in shape tables. (See *AISC Commentary* Fig. C-D3.1 for guidance on establishing $\overline{x}$.) l is the length of the connection in the direction of loading, measured for bolted connections as the center-to-center distance between the two end connectors in a line.

Table 60.2 Shear Lag Reduction Coefficient (U) Values for Welded Connections

shape	condition	U
all tension members where load is transmitted across each cross-sectional element by welds	all	1.00
W, M, S, and HP, and structural tees cut from them, where load is transmitted across some but not all of cross-sectional elements by *longitudinal* welds		$U = (1 - \overline{x})/l$ l is the longitudinal weld length. $\overline{x}$ is as defined in Table 60.1.
all tension members except plates and HSS, where load is transmitted across some but not all of cross-sectional elements by *transverse* welds		1.00 A_n is the sum of areas of only the directly connected cross-sectional elements
plate where load is transmitted by *longitudinal* welds (w is plate width perpendicular to load; l is longitudinal weld length)	$l \geq 2w$	1.00
	$2w > l \geq 1.5w$	0.87
	$1.5w > l \geq w$	0.75
round or rectangular HSS sections		See *AISC Specification* Table D3.1

3. EFFECTIVE LENGTH

$$F_e = \frac{\pi^2 E}{\left(\frac{KL}{r}\right)^2} \quad [\textit{AISC Specification} \text{ Eq. E3-4}] \qquad 61.3$$

$$G = \frac{\sum\left(\frac{I}{L}\right)_c}{\sum\left(\frac{I}{L}\right)_b} \qquad 61.4$$

Table 61.1 Effective Length Factors*

end no. 1	end no. 2	K
built-in	built-in	0.65
built-in	pinned	0.80
built-in	rotation fixed, translation free	1.2
built-in	free	2.1
pinned	pinned	1.0
pinned	rotation fixed, translation free	2.0

*These are slightly different from the theoretical values often quoted for use with Euler's equation.

5. SLENDERNESS RATIO

$$SR = \frac{KL}{r} \qquad 61.5$$

7. DESIGN COMPRESSIVE STRENGTH

$$F_{cr} = (0.658^{F_y/F_e})F_y \qquad 61.6$$

$$F_{cr} = 0.877F_e \qquad 61.7$$

Figure 61.4 Available Compressive Stress Versus Slenderness Ratio

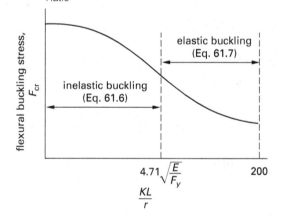

8. ANALYSIS OF COLUMNS

$$P_n = F_{cr}A_g \quad [\textit{AISC Specification} \text{ Eq. E3-1}] \qquad 61.8$$

$$SR = \text{larger of} \begin{Bmatrix} \dfrac{K_x L_x}{r_x} \\ \dfrac{K_y L_y}{r_y} \end{Bmatrix} \qquad 61.9$$

$$P_a = \frac{P_n}{\Omega_c} = \frac{F_{cr}A_g}{1.67} \qquad [\text{ASD}] \quad 61.10(a)$$

$$\phi_c P_n = 0.90 F_{cr}A_g \qquad [\text{LRFD}] \quad 61.10(b)$$

9. DESIGN OF COLUMNS

$$\text{effective length} = K_y L_y \qquad \textit{61.11}$$

$$K_x L_x' = \frac{K_x L_x}{\dfrac{r_x}{r_y}} \qquad \textit{61.12}$$

10. LOCAL BUCKLING

(See Table 61.2.)

$$\frac{b}{t} \le \lambda_p \quad \text{[compact]} \qquad \textit{61.13}$$

$$\lambda_p < \frac{b}{t} \le \lambda_r \quad \text{[noncompact]} \qquad \textit{61.14}$$

12. COLUMN BASE PLATES

$$P_a = \frac{P_p}{\Omega_c} \quad [\text{ASD}; \ \Omega_c = 2.50] \qquad \textit{61.16(a)}$$

$$P_u = \phi_c P_p \quad [\text{LRFD}; \ \phi_c = 0.60] \qquad \textit{61.16(b)}$$

$$A_1 = \frac{\text{unfactored column load}}{F_p} \qquad \textit{61.17}$$

$$F_p = 0.85 f_c' \sqrt{\frac{A_2}{A_1}} \le 1.7 f_c' \qquad \textit{61.18}$$

$$P_p = F_p A_1 = 0.85 f_c' A_1 \sqrt{\frac{A_2}{A_1}} \le 1.7 f_c' A_1 \qquad \textit{61.19}$$

$$N = 2m + 0.95d \qquad \textit{61.20}$$

$$B = \frac{A_1}{N} \qquad \textit{61.21}$$

$$f_p = \frac{\text{unfactored column load}}{BN} \qquad \textit{61.22}$$

$$m = \frac{N - 0.95d}{2} \qquad \textit{61.23}$$

$$n = \frac{B - 0.80 b_f}{2} \qquad \textit{61.24}$$

$$M_n = F_y S \quad [\text{ASD}] \qquad \textit{61.25(a)}$$

$$M_n = F_y Z \quad [\text{LRFD}] \qquad \textit{61.25(b)}$$

$$Z = \frac{\{N \text{ or } B\} t^2}{4} \qquad \textit{61.26}$$

$$S = \frac{\{N \text{ or } B\} t^2}{6} \qquad \textit{61.27}$$

$$n' = \frac{\sqrt{db_f}}{4} \qquad \textit{61.28}$$

$$\lambda = \frac{2\sqrt{X}}{1 + \sqrt{1 - X}} \le 1.0 \qquad \textit{61.29}$$

$$X = \left(\frac{4 d h_f}{(d + b_f)^2}\right)\left(\frac{P_{u,\text{req'd}}}{\phi_c P_p}\right) \quad [\text{LRFD}; \ \phi_c = 0.6] \qquad \textit{61.30(a)}$$

$$X = \left(\frac{4 d b_f}{(d + b_f)^2}\right)\left(\frac{\Omega_u P_{a,\text{req'd}}}{P_p}\right) \quad [\text{ASD}; \ \Omega_c = 2.50] \qquad \textit{61.30(b)}$$

$$M_n = \tfrac{1}{2} l^2 \{B \text{ or } N\} f_p \qquad \textit{61.31}$$

Table 61.2 *Limiting Width-to-Thickness Ratios for Elements in Axial Compression*

element	ratio	λ_r (nonslender/slender)
flanges of rolled I-shaped sections; plates projecting from rolled I-shaped sections; outstanding legs of pairs of angles connected with continuous contact; flanges of channels; and flanges of tees	b/t	$0.56\sqrt{E/F_y}$
flanges of built-up I-shaped sections and plates or angle legs projecting from built-up I-shaped sections	b/t	$0.64\sqrt{k_c E/F_y}$ *
legs of single angles, legs of double angles with separators, and all other unstiffened elements	b/t	$0.45\sqrt{E/F_y}$
stems of tees	d/t	$0.75\sqrt{E/F_y}$
webs of doubly symmetric I-shaped sections and channels	h/t_w	$1.49\sqrt{E/F_y}$
walls of rectangular HSS and boxes of uniform thickness	b/t	$1.40\sqrt{E/F_y}$
flange cover plates and diaphragm plates between lines of fasteners or welds	b/t	$1.40\sqrt{E/F_y}$
all other stiffened elements	b/t	$1.49\sqrt{E/F_y}$
round HSS	D/t	$0.11 E/F_y$

*$k_c = 4/\sqrt{h/t_w}$; but, $0.35 \le k_c \le 0.76$.

Source: Adapted from *AISC Specification* Table B4.1a.

Figure 61.7 *Column Base Plate*

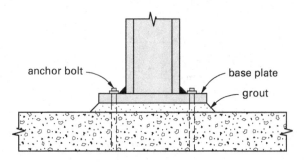

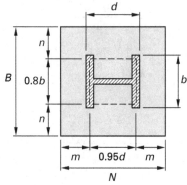

$$t_{\min} = l\sqrt{\frac{2P_u}{0.9F_yBN}} \qquad \text{[LRFD]} \quad 61.32(a)$$

$$t_{\min} = l\sqrt{\frac{3.33P_a}{F_yBN}} \qquad \text{[ASD]} \quad 61.32(b)$$

CERM Chapter 62
Structural Steel: Beam-Columns

1. INTRODUCTION

$$f_{\max} = f_a + (\text{AF})_x f_{bx} + (\text{AF})_y f_{by} \qquad 62.1$$

$$\frac{f_a}{F_a} + (\text{AF})_x\left(\frac{f_{bx}}{F_{bx}}\right) + (\text{AF})_y\left(\frac{f_{by}}{F_{by}}\right) = 1.0 \qquad 62.2$$

2. FLEXURAL/AXIAL COMPRESSION

For $P_r/P_c \geq 0.2$, AISC Eq. H1-1a is

$$\frac{P_r}{P_c} + \left(\tfrac{8}{9}\right)\left(\frac{M_{rx}}{M_{cx}} + \frac{M_{ry}}{M_{cy}}\right) \leq 1.0 \qquad 62.3$$

For $P_r/P_c < 0.2$, AISC Eq. H1-1b is

$$\frac{P_r}{2P_c} + \frac{M_{rx}}{M_{cx}} + \frac{M_{ry}}{M_{cy}} \leq 1.0 \qquad 62.4$$

$$\frac{P_r}{P_{cy}}\left(1.5 - 0.5\frac{P_r}{P_{cy}}\right) + \left(\frac{M_{rx}}{C_b M_{cx}}\right)^2 \leq 1.0 \qquad 62.5$$

[AISC Eq. H1-2]

$$C_b = \frac{12.5M_{\max}}{2.5M_{\max} + 3M_A + 4M_B + 3M_C} \qquad 62.6$$

[AISC Eq. F1-1]

3. SECOND-ORDER EFFECTS

$$M_r = B_1 M_{nt} + B_2 M_{lt} \quad \text{[AISC Eq. A-8-1]} \qquad 62.7$$

$$B_1 = \frac{C_m}{1 - \dfrac{\alpha P_r}{P_{e1}}} \geq 1 \quad \text{[AISC Eq. A-8-3]} \qquad 62.8$$

$$B_2 = \frac{1}{1 - \dfrac{\alpha P_{\text{story}}}{P_{e,\text{story}}}} \geq 1 \quad \text{[AISC Eq. A-8-6]} \qquad 62.9$$

5. DESIGN OF BEAM-COLUMNS

For $P_r/P_c \geq 0.2$,

$$pP_r + b_x M_{rx} + b_y M_{ry} \leq 1.0 \quad \text{[large axial loads]} \qquad 62.10$$

For $P_r/P_c < 0.2$,

$$\frac{pP_r}{2} + \left(\tfrac{9}{8}\right)(b_x M_{rx} + b_y M_{ry}) \leq 1.0 \quad \text{[small axial loads]} \qquad 62.11$$

CERM Chapter 63
Structural Steel: Built-Up Sections

1. INTRODUCTION

Figure 63.1 *Elements of a Built-Up Section*

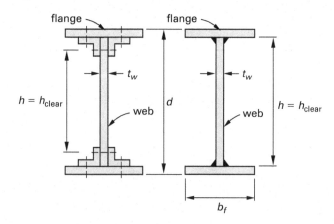

2. DEPTH-THICKNESS RATIOS

$$\left(\frac{h}{t_w}\right)_{\max} \leq 12.0\sqrt{\frac{E}{F_y}} \qquad 63.1$$

$$\left(\frac{h}{t_w}\right)_{\max} \leq \frac{0.40E}{F_y} \qquad 63.2$$

3. SHEAR STRENGTH

$$V_n = 0.6 F_y A_w C_v \quad [AISC \ Specification \ \text{Eq. G2-1}] \qquad 63.3$$

For $h/t_w \le 1.10\sqrt{k_v E/F_y}$,

$$C_v = 1.0 \quad [AISC \ Specification \ \text{Eq. G2-3}] \qquad 63.4$$

For $1.10\sqrt{k_v E/F_y} < h/t_w \le 1.37\sqrt{k_v E/F_y}$,

$$C_v = \frac{1.10\sqrt{\dfrac{k_v E}{F_y}}}{\dfrac{h}{t_w}} \quad [AISC \ Specification \ \text{Eq. G2-4}] \qquad 63.5$$

For $h/t_w > 1.37\sqrt{k_v E/F_y}$,

$$C_v = \frac{1.51 k_v E}{\left(\dfrac{h}{t_w}\right)^2 F_y} \quad [AISC \ Specification \ \text{Eq. G2-5}] \qquad 63.6$$

4. DESIGN OF GIRDER WEBS AND FLANGES

$$Z_x = A_f \left(h + t_f\right) + \frac{t_w h^2}{4} \qquad 63.7$$

$$A_f = \frac{M_p}{F_y h} - \frac{ht}{4} \qquad 63.8$$

$$A_f = b_f t_f \qquad 63.9$$

5. WIDTH-THICKNESS RATIOS

$$\frac{b_f}{2 t_f} \le \frac{64.7}{\sqrt{F_{yf}}} \qquad 63.10$$

$$\frac{b_f}{2 t_f} \le 162\sqrt{\frac{k_c}{F_L}} \qquad 63.11$$

8. LOCATION OF INTERIOR STIFFENERS

$$k_v = 5 + \frac{5}{\left(\dfrac{a}{h}\right)^2} \qquad 63.12$$

$k_v = 5$ applies when $a/h > 3.0$, or

$$\frac{a}{h} > \left(\frac{260}{\dfrac{h}{t_w}}\right)^2 \qquad 63.13$$

For $h/t_w > 1.10\sqrt{k_v E/F_y}$,

$$V_n = 0.6 F_y A_w \left(C_v + \frac{1 - C_v}{1.15\sqrt{1 + \left(\dfrac{a}{h}\right)^2}} \right)$$

$$[AISC \ Specification \ \text{Eq. G3-2}] \qquad 63.14$$

For $h/t_w \le 1.10\sqrt{k_v E/F_y}$,

$$V_n = 0.6 F_y A_w \qquad 63.15$$

9. DESIGN OF INTERMEDIATE STIFFENERS

$$I_{st} \ge I_{st1} + (I_{st2} - I_{st1})\left(\frac{V_r - V_{c1}}{V_{c2} - V_{c1}}\right) \qquad 63.16$$

$$\left(\frac{b}{t}\right)_{st} \le 0.56\sqrt{\frac{E}{F_{y,st}}} \qquad 63.17$$

Figure 63.2 Intermediate Stiffeners

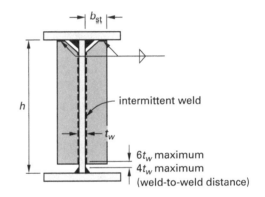

10. DESIGN OF BEARING STIFFENERS

$$\frac{b_{st}}{t_{st}} \le \frac{95}{\sqrt{F_y}} \qquad 63.18$$

$$\frac{l}{r} = \frac{0.75h}{0.25(2b_{st} + t_w)} \qquad 63.19$$

$$t_{st} = \frac{\dfrac{\text{load}}{\phi F_{cr}} - 25 t_w^2}{2 b_{st}} \qquad 63.20$$

$$t_{st} = \frac{\dfrac{\text{load}}{0.90 F_y}}{2 b_{st}} \qquad 63.21$$

Figure 63.3 Bearing Stiffener (top view)

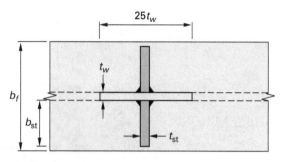

CERM Chapter 64
Structural Steel: Composite Beams

2. EFFECTIVE WIDTH OF CONCRETE SLAB

$$b = \text{smaller of} \begin{cases} \dfrac{L}{4} & \text{[interior beams]} \\ s \end{cases}$$ 64.1

3. SECTION PROPERTIES

$$I_{\text{LB}} = I_s + A_s(Y_{\text{ENA}} - d_3)^2$$
$$+ \left(\frac{\sum Q_n}{F_y}\right)(2d_3 + d_1 - Y_{\text{ENA}})^2$$

[AISC Commentary Eq. C-I3-1] 64.2

$$Y_{\text{ENA}} = \frac{A_s d_3 + \left(\dfrac{\sum Q_n}{F_y}\right)(2d_3 + d_1)}{A_s + \dfrac{\sum Q_n}{F_y}}$$

Figure 64.2 Deflection Design Model for Composite Beams (AISC Manual Fig. 3-4)*

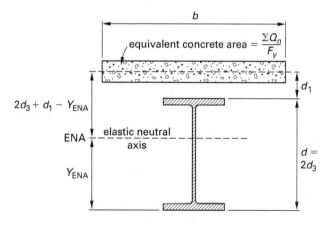

AISC Manual Fig. 3-4 uses Y2 for d_1 and $d + Y2 - Y_{\text{ENA}}$ for $2d_3 + d_1 - Y_{\text{ENA}}$.

4. AVAILABLE FLEXURAL STRENGTH

$$M_n = A_g F_y\left(\frac{d}{2} + Y2\right)$$ 64.3

5. SHEAR STUD CONNECTORS

$$N = \frac{\sum Q_n}{Q_n}$$ 64.4

CERM Chapter 65
Structural Steel: Connectors

4. AVAILABLE LOADS FOR FASTENERS

Slip resistance, ϕR_n or $\dfrac{R_n}{\Omega}$:

$$R_n = \mu D_u h_f T_b n_s$$ 65.1

$$T_{b,\min} = 0.70 A_{b,\text{th}} F_u \quad \text{[rounded]}$$ 65.2

Table 65.1 Nominal Fastener Stresses for Static Loading

type of connector	F_{nt} (ksi)	F_{nv} in bearing connections (ksi)
A307 common bolts	45	27
A325 high-strength bolts		
no threads in shear plane	90	68
threads in shear plane	90	54
A490 high-strength bolts		
no threads in shear plane	113	84
threads in shear plane	113	68

Source: Based on *AISC Specification* Table J3.2.

5. AVAILABLE BEARING STRENGTH

$$R_n = F_n A_b$$ 65.3

8. ULTIMATE STRENGTH OF ECCENTRIC SHEAR CONNECTIONS

$$P = C R_n$$ 65.4

9. TENSION CONNECTIONS

$$P = A_b F_{nt}$$ 65.5

10. COMBINED SHEAR AND TENSION CONNECTIONS

Figure 65.4 *Combined Shear and Tension Connection*

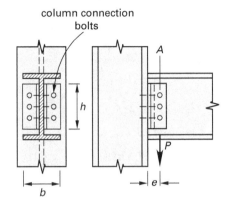

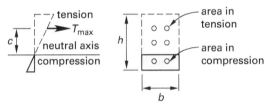

CERM Chapter 66
Structural Steel: Welding

3. FILLET WELDS

(See Table 66.1.)

$$t_e = 0.707w \qquad 66.1$$

$$F_v = 0.60F_{u,\text{rod}} \qquad 66.2$$

$$f_v = \frac{P}{A_w} = \frac{P}{l_w t_e} \qquad 66.3$$

$$R_w = 0.30t_e F_{u,\text{rod}} \qquad 66.4$$

Figure 66.4 *Fillet Weld*

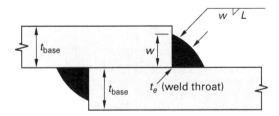

4. CONCENTRIC TENSION CONNECTIONS

$$\phi R_n = 1.392Dl \quad \text{[LRFD; E70 electrode]} \qquad 66.5$$

$$\frac{R_n}{\Omega} = 0.928Dl \quad \text{[ASD; E70 electrode]} \qquad 66.6$$

Table 66.1 *Minimum Fillet Weld Size*

thickness of thinner part joined (in)	minimum weld size* w (in)
to $\frac{1}{4}$ inclusive	$\frac{1}{8}$
over $\frac{1}{4}$ to $\frac{1}{2}$ inclusive	$\frac{3}{16}$
over $\frac{1}{2}$ to $\frac{3}{4}$ inclusive	$\frac{1}{4}$
over $\frac{3}{4}$	$\frac{5}{16}$

(Multiply in by 25.4 to obtain mm.)
*Leg dimension of fillet weld. Single pass welds must be used.

Source: *AISC Specification* Table J2.4

6. ECCENTRICALLY LOADED WELDED CONNECTIONS

Figure 66.7 *Welded Connection in Combined Shear and Bending (eccentricity normal to plane of faying surfaces)*

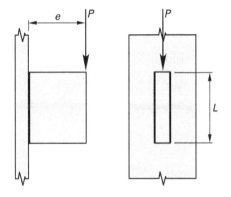

7. ELASTIC METHOD FOR ECCENTRIC SHEAR/ TORSION CONNECTIONS

$$I_p = I_x + I_y \qquad 66.12$$

$$r_m = \frac{Mc}{I_p} = \frac{Pec}{I_p} \qquad 66.13$$

$$r_p = \frac{P}{A} \qquad 66.14$$

$$r_a = \sqrt{(r_{m,y} + r_p)^2 + (r_{m,x})^2} \qquad 66.15$$

8. AISC INSTANTANEOUS CENTER OF ROTATION METHOD FOR ECCENTRIC SHEAR/TORSION CONNECTIONS

$$R_n = CC_1 Dl \qquad 66.16$$

9. ELASTIC METHOD FOR OUT OF PLANE LOADING

$$f_v = \frac{P_n}{A_w} = \frac{P_n}{2Lt_e} \qquad 66.17$$

$$f_b = \frac{Mc}{I} = \frac{M}{S} = \frac{P_n e}{S} \qquad 66.18$$

$$f = \sqrt{f_v^2 + f_b^2} \qquad 66.19$$

CERM Chapter 67
Properties of Masonry
(Only the ASD methods presented in *MSJC* may be used on the exam.)

8. MODULUS OF ELASTICITY

$$E_m = 700 f'_m \quad \text{[clay masonry]} \qquad 67.1$$

$$E_m = 900 f'_m \quad \text{[concrete masonry]} \qquad 67.2$$

18. DEVELOPMENT LENGTH

For wires in tension,

$$l_d = 0.22 d_b F_s \qquad \text{[SI]} \quad 67.3(a)$$

$$l_d = 0.0015 d_b F_s \qquad \text{[U.S.]} \quad 67.3(b)$$

For reinforcing bars,

$$l_d = \frac{1.57 d_b^2 f_y \gamma}{K \sqrt{f'_m}} \qquad \text{[SI]} \quad 67.4(a)$$

$$l_d = \frac{0.13 d_b^2 f_y \gamma}{K \sqrt{f'_m}} \qquad \text{[U.S.]} \quad 67.4(b)$$

Increase lengths by 50% for epoxy-coated bars and wires.

For ASD,

$$l_d = 0.29 d_b f_s \qquad \text{[SI]} \quad 67.5(a)$$

$$l_d = 0.002 d_b f_s \qquad \text{[U.S.]} \quad 67.5(b)$$

CERM Chapter 68
Masonry Walls
(Only the ASD methods presented in *MSJC* may be used on the exam; except for strength design (SD) Sec. 9.3.5, which may be used for walls with out-of-plane loads.)

8. ASD WALL DESIGN: FLEXURE—REINFORCED

$$F_b = 0.45 f'_m \qquad 68.1$$

$$F_v = F_{vm} + F_{vs} \qquad 68.2$$

$$\rho = \frac{A_s}{bd} \qquad 68.3$$

$$k = \sqrt{2\rho n + (\rho n)^2} - \rho n \qquad 68.4$$

$$j = 1 - \frac{k}{3} \qquad 68.5$$

$$M_m = F_b b d^2 \left(\frac{jk}{2} \right) \qquad 68.6$$

$$M_s = A_s F_s j d \qquad 68.7$$

$$M_R = \text{the lesser of } M_m \text{ and } M_s \qquad 68.8$$

$$V_R = F_v b d \qquad 68.9$$

$$\rho_{\text{bal}} = \frac{n F_b}{2 F_s \left(n + \dfrac{F_s}{F_b} \right)} \quad \text{[balanced]} \qquad 68.10$$

$$k = \frac{-A_s n - t_{\text{fs}}(b - b_w)}{d b_w}$$

$$+ \frac{\sqrt{\begin{array}{c} \left(A_s n + t_{\text{fs}}(b - b_w) \right)^2 \\ + t_{\text{fs}}^2 b_w (b - b_w) + 2 d b_w A_s n \end{array}}}{d b_w} \qquad 68.11$$

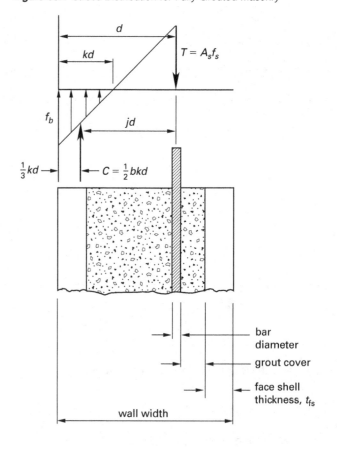

Figure 68.1 *Stress Distribution for Fully Grouted Masonry*

$$j = \left(\frac{1}{kdb_w + t_{\text{fs}}(b - b_w)\left(2 - \frac{t_{\text{fs}}}{kd}\right)} \right)$$
$$\times \left(\begin{array}{c} kb_w\left(d - \frac{kd}{3}\right) + \left(\frac{2t_{\text{fs}}(b - b_w)}{kd^2}\right) \\ \times (kd - t_{\text{fs}})\left(d - \frac{t_{\text{fs}}}{2}\right) + \left(\frac{t_{\text{fs}}}{2}\right)\left(d - \frac{t_{\text{fs}}}{3}\right) \end{array} \right)$$

68.12

$$M_m = \tfrac{1}{2}F_b kdb_w\left(d - \frac{kd}{3}\right) + F_b t_{\text{fs}}(b - b_w)$$
$$\times \left(\left(1 - \frac{t_{\text{fs}}}{kd}\right)\left(d - \frac{t_{\text{fs}}}{2}\right) + \left(\frac{t_{\text{fs}}}{2kd}\right)\left(d - \frac{t_{\text{fs}}}{3}\right) \right)$$

68.13

$$M_s = A_s F_s jd \qquad\qquad 68.14$$

$$M_R = \text{the lesser of } M_m \text{ and } M_s \qquad 68.15$$

$$V_R = F_v\left(bt_{\text{fs}} + b_w(d - t_{\text{fs}})\right) \qquad 68.16$$

Table 68.3 *Reinforcing Steel Areas, A_s, in in^2/ft*

bar spacing (in)	reinforcing bar size						
	no. 3	no. 4	no. 5	no. 6	no. 7	no. 8	no. 9
8	0.166	0.295	0.460	0.663	0.902	1.178	1.491
16	0.083	0.147	0.230	0.331	0.451	0.589	0.746
24	0.055	0.098	0.153	0.221	0.301	0.393	0.497
32	0.041	0.074	0.115	0.166	0.225	0.295	0.373
40	0.033	0.059	0.092	0.133	0.180	0.236	0.298
48	0.028	0.049	0.077	0.110	0.150	0.196	0.249
56	0.024	0.042	0.066	0.095	0.129	0.168	0.213
64	0.021	0.037	0.058	0.083	0.113	0.147	0.186
72	0.018	0.033	0.051	0.074	0.100	0.131	0.166

9. ASD WALL DESIGN: AXIAL COMPRESSION AND FLEXURE—UNREINFORCED

$$\frac{f_a}{F_a} + \frac{f_b}{F_b} \leq 1 \qquad 68.17$$

$$P \leq \tfrac{1}{4}P_e \qquad 68.18$$

$$P_e = \left(\frac{\pi^2 E_m I_n}{h^2}\right)\left(1 - 0.577\left(\frac{e}{r}\right)\right)^3 \qquad 68.19$$

$$F_a = \tfrac{1}{4}f'_m\left(1 - \left(\frac{h}{140r}\right)^2\right) \quad [\text{for } h/r \leq 99] \qquad 68.20$$

$$F_a = \tfrac{1}{4}f'_m\left(\frac{70r}{h}\right)^2 \quad [\text{for } h/r > 99] \qquad 68.21$$

$$F_b = \tfrac{1}{3}f'_m \qquad 68.22$$

11. ASD WALL DESIGN: AXIAL COMPRESSION AND FLEXURE—REINFORCED

$$F_b = 0.45f'_m \qquad 68.26$$

$$P_a = (0.25f'_m A_n + 0.65A_s F_s)\left(1 - \left(\frac{h}{140r}\right)^2\right)$$
$$[\text{when } h/r \leq 99] \qquad 68.27$$

$$P_a = (0.25f'_m A_n + 0.65A_s F_s)\left(\frac{70r}{h}\right)^2$$
$$[\text{when } h/r > 99] \qquad 68.28$$

$$k_b = \frac{F_b}{F_b + \dfrac{F_s}{n}} \qquad 68.29$$

12. SD WALL DESIGN: AXIAL COMPRESSION AND FLEXURE—REINFORCED

From *MSJC* Sec. 9.3.5,

$$M_u = \frac{w_u h^2}{8} + P_{uf}\frac{e_u}{2} + P_u \delta_u \qquad 68.30$$

$$M_u \leq \phi M_n \qquad 68.31$$

$$M_n = (A_s f_y + P_u)\left(d - \frac{a}{2}\right) \qquad 68.32$$

$$a = \frac{A_s f_y + P_u}{0.80f'_m b} \qquad 68.33$$

$$M_{\text{cr}} = Sf_r \qquad 68.34$$

$$\delta_s = \frac{5M_{\text{ser}}h^2}{48E_m I_g} \qquad 68.35$$

$$\delta_s = \frac{5M_{\text{cr}}h^2}{48E_m I_g} + \frac{5(M_{\text{ser}} - M_{\text{cr}})h^2}{48E_m I_{\text{cr}}} \qquad 68.36$$

14. ASD WALL DESIGN: SHEAR WALLS WITH NO NET TENSION—UNREINFORCED

$$f_v = \frac{VQ}{I_n b} = \frac{3V}{2A_n} \quad [\text{rectangular sections}] \qquad 68.37$$

Under *MSJC* Sec. 8.2.6.2, allowable in-plane shear stress, F_v, is the least of

(a) $1.5\sqrt{f'_m}$ $\qquad\qquad\qquad\qquad\qquad$ *68.38*

(b) 120 lbf/in² (0.83 MPa) $\qquad\qquad$ *68.39*

(c) $v + 0.45N_v/A_n$ $\qquad\qquad\qquad\quad$ *68.40*

16. ASD WALL DESIGN: SHEAR WALLS WITH NET TENSION—REINFORCED

$$f_v = \frac{V}{bd} = \frac{V}{A_{nv}} \qquad 68.46$$

$$F_v \le \left(3\sqrt{f'_m}\right)\gamma_g \quad [M/(Vd) \le 0.25] \qquad 68.47(a)$$

$$F_v \le \left(2\sqrt{f'_m}\right)\gamma_g \quad [M/(Vd) \le 1.0] \qquad 68.47(b)$$

$$F_{vm} = \frac{1}{2}\left(\left(4 - 1.75\left(\frac{M}{vd}\right)\right)\sqrt{f'_m}\right) + 0.25\frac{P}{A_n}$$

[for all masonry except special reinforced shear walls]
$$68.48$$

$$F_{vm} = \frac{1}{4}\left(\left(4 - 1.75\left(\frac{M}{vd}\right)\right)\sqrt{f'_m}\right) + 0.25\frac{P}{A_n}$$

[for special reinforced masonry shear walls]
$$68.49$$

$$F_{vs} = \frac{1}{2}\left(\frac{A_v F_s d}{A_n s}\right) \qquad 68.50$$

CERM Chapter 69
Masonry Columns
(Only the ASD methods presented in *MSJC* may be used on the exam.)

5. ASD DESIGN FOR PURE COMPRESSION

$$P_a = (0.25f'_m A_n + 0.65A_{st}F_s)\left(1 - \left(\frac{h}{140r}\right)^2\right)$$

[when $h/r \le 99$]   69.1

$$P_a = (0.25f'_m A_n + 0.65A_{st}F_s)\left(\frac{70r}{h}\right)^2$$

[when $h/r > 99$]   69.2

9. BIAXIAL BENDING

$$\frac{1}{P_{\text{biaxial}}} = \frac{1}{P_x} + \frac{1}{P_y} - \frac{1}{P_o} \qquad 69.6$$

Transportation

CERM Chapter 70
Properties of Solid Bodies

1. CENTER OF GRAVITY

$$x_c = \frac{\int x \, dm}{m} \qquad 70.1$$

$$y_c = \frac{\int y \, dm}{m} \qquad 70.2$$

$$z_c = \frac{\int z \, dm}{m} \qquad 70.3$$

$$x_c = \frac{\sum m_i x_{ci}}{\sum m_i} \qquad 70.4$$

$$y_c = \frac{\sum m_i y_{ci}}{\sum m_i} \qquad 70.5$$

$$z_c = \frac{\sum m_i z_{ci}}{\sum m_i} \qquad 70.6$$

2. MASS AND WEIGHT

$$m = \rho V \qquad 70.7$$

$$w = mg \qquad \text{[SI]} \quad 70.8(a)$$

$$w = \frac{mg}{g_c} \qquad \text{[U.S.]} \quad 70.8(b)$$

4. MASS MOMENT OF INERTIA

$$I_x = \int (y^2 + z^2) \, dm \qquad 70.9$$

$$I_y = \int (x^2 + z^2) \, dm \qquad 70.10$$

$$I_z = \int (x^2 + y^2) \, dm \qquad 70.11$$

5. PARALLEL AXIS THEOREM

$$I_{\text{any parallel axis}} = I_c + md^2 \qquad 70.12$$

$$I = I_{c,1} + m_1 d_1^2 + I_{c,2} + m_2 d_2^2 + \cdots \qquad 70.13$$

6. RADIUS OF GYRATION

$$k = \sqrt{\frac{I}{m}} \qquad 70.14$$

$$I = k^2 m \qquad 70.15$$

CERM Chapter 71
Kinematics

5. LINEAR PARTICLE MOTION

$$s(t) = \int \mathrm{v}(t) \, dt = \int \left(\int a(t) \, dt \right) dt \qquad 71.2$$

$$\mathrm{v}(t) = \frac{ds(t)}{dt} = \int a(t) \, dt \qquad 71.3$$

$$a(t) = \frac{d\mathrm{v}(t)}{dt} = \frac{d^2 s(t)}{dt^2} \qquad 71.4$$

$$\mathrm{v}_{\text{ave}} = \frac{\int_1^2 \mathrm{v}(t) \, dt}{t_2 - t_1} = \frac{s_2 - s_1}{t_2 - t_1} \qquad 71.5$$

$$a_{\text{ave}} = \frac{\int_1^2 a(t) \, dt}{t_2 - t_1} = \frac{\mathrm{v}_2 - \mathrm{v}_1}{t_2 - t_1} \qquad 71.6$$

6. DISTANCE AND SPEED

$$\text{displacement} = s(t_2) - s(t_1) \qquad 71.7$$

7. UNIFORM MOTION

$$s(t) = s_0 + vt \qquad \text{71.8}$$

$$v(t) = v \qquad \text{71.9}$$

$$a(t) = 0 \qquad \text{71.10}$$

8. UNIFORM ACCELERATION

$$a(t) = a \qquad \text{71.11}$$

$$v(t) = a\int dt = v_0 + at \qquad \text{71.12}$$

$$s(t) = a\iint dt^2 = s_0 + v_0 t + \tfrac{1}{2}at^2 \qquad \text{71.13}$$

Table 71.1 Uniform Acceleration Formulas*

to find	given these	use this equation
a	t, v_0, v	$a = \dfrac{v - v_0}{t}$
a	t, v_0, s	$a = \dfrac{2s - 2v_0 t}{t^2}$
a	v_0, v, s	$a = \dfrac{v^2 - v_0^2}{2s}$
s	t, a, v_0	$s = v_0 t + \tfrac{1}{2}at^2$
s	a, v_0, v	$s = \dfrac{v^2 - v_0^2}{2a}$
s	t, v_0, v	$s = \tfrac{1}{2}t(v_0 + v)$
t	a, v_0, v	$t = \dfrac{v - v_0}{a}$
t	a, v_0, s	$t = \dfrac{\sqrt{v_0^2 + 2as} - v_0}{a}$
t	v_0, v, s	$t = \dfrac{2s}{v_0 + v}$
v_0	t, a, v	$v_0 = v - at$
v_0	t, a, s	$v_0 = \dfrac{s}{t} - \tfrac{1}{2}at$
v_0	a, v, s	$v_0 = \sqrt{v^2 - 2as}$
v	t, a, v_0	$v = v_0 + at$
v	a, v_0, s	$v = \sqrt{v_0^2 + 2as}$

*The table can be used for rotational problems by substituting α, ω, and θ for a, v, and s, respectively.

CERM Chapter 72
Kinetics

5. LINEAR MOMENTUM

$$\mathbf{p} = m\mathbf{v} \qquad \text{[SI]} \quad \text{72.1(a)}$$

$$\mathbf{p} = \dfrac{m\mathbf{v}}{g_c} \qquad \text{[U.S.]} \quad \text{72.1(b)}$$

9. NEWTON'S SECOND LAW OF MOTION

$$F = m\left(\dfrac{dv}{dt}\right) = ma \qquad \text{[SI]} \quad \text{72.12(a)}$$

$$F = \left(\dfrac{m}{g_c}\right)\left(\dfrac{dv}{dt}\right) = \dfrac{ma}{g_c} \qquad \text{[U.S.]} \quad \text{72.12(b)}$$

$$T = I\left(\dfrac{d\omega}{dt}\right) = I\alpha \qquad \text{[SI]} \quad \text{72.14(a)}$$

$$T = \left(\dfrac{I}{g_c}\right)\left(\dfrac{d\omega}{dt}\right) = \dfrac{I\alpha}{g_c} \qquad \text{[U.S.]} \quad \text{72.14(b)}$$

10. CENTRIPETAL FORCE

$$F_c = ma_n = \dfrac{mv_t^2}{r} \qquad \text{[SI]} \quad \text{72.15(a)}$$

$$F_c = \dfrac{ma_n}{g_c} = \dfrac{mv_t^2}{g_c r} \qquad \text{[U.S.]} \quad \text{72.15(b)}$$

13. FLAT FRICTION

$$N = mg \qquad \text{[SI]} \quad \text{72.19(a)}$$

$$N = \dfrac{mg}{g_c} \qquad \text{[U.S.]} \quad \text{72.19(b)}$$

$$N = mg\cos\phi \qquad \text{[SI]} \quad \text{72.20(a)}$$

$$N = \dfrac{mg\cos\phi}{g_c} \qquad \text{[U.S.]} \quad \text{72.20(b)}$$

$$F_{f,\text{max}} = f_s N \qquad \text{72.21}$$

$$\tan\phi = f_s \qquad \text{72.22}$$

16. ROLLING RESISTANCE

$$F_r = \dfrac{mga}{r} \qquad \text{[SI]} \quad \text{72.27(a)}$$

$$F_r = \dfrac{mga}{rg_c} = \dfrac{wa}{r} \qquad \text{[U.S.]} \quad \text{72.27(b)}$$

$$f_r = \dfrac{F_r}{w} = \dfrac{a}{r} \qquad \text{72.28}$$

Figure 72.11 *Wheel Rolling Resistance*

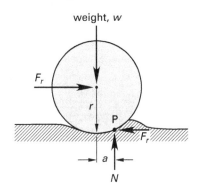

17. ROADWAY/CONVEYOR BANKING

$$\tan \phi = \frac{v_t^2}{gr} \qquad 72.29$$

$$v_t = \sqrt{gr \tan \phi} \qquad 72.30$$

$$e = \tan \phi = \frac{v_t^2 - fgr}{gr + fv_t^2} \qquad 72.31$$

Figure 72.12 *Roadway Banking*

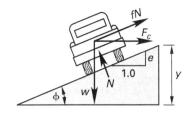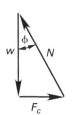

- For small banking angles (i.e., $\phi \leq 8°$),

$$r \approx \frac{v_t^2}{g(e + f)} \qquad 72.32$$

$$r_\mathrm{m} = \frac{v_{\mathrm{km/h}}^2}{127(e_{\mathrm{m/m}} + f)} \qquad [\text{SI}] \qquad 72.33(a)$$

$$r_\mathrm{ft} = \frac{v_{\mathrm{mph}}^2}{15(e_{\mathrm{ft/ft}} + f)} \qquad [\text{U.S.}] \qquad 72.33(b)$$

18. MOTION OF RIGID BODIES

$$\sum F_x = ma_x \qquad [\text{consistent units}] \qquad 72.34$$

$$\sum F_y = ma_y \qquad [\text{consistent units}] \qquad 72.35$$

$$T = I\alpha \qquad [\text{SI}] \qquad 72.36(a)$$

$$T = \frac{I\alpha}{g_c} \qquad [\text{U.S.}] \qquad 72.36(b)$$

21. IMPULSE

$$\text{Imp} = \int_{t_1}^{t_2} F \, dt \quad [\text{linear}] \qquad 72.49$$

$$\text{Imp} = \int_{t_1}^{t_2} T \, dt \quad [\text{angular}] \qquad 72.50$$

$$\text{Imp} = F(t_2 - t_1) \quad [\text{linear}] \qquad 72.51$$

$$\text{Imp} = T(t_2 - t_1) \quad [\text{angular}] \qquad 72.52$$

$$F_{\text{ave}} = \frac{\text{Imp}}{\Delta t} \qquad 72.53$$

22. IMPULSE-MOMENTUM PRINCIPLE

$$\text{Imp} = \Delta p \qquad 72.54$$

$$F(t_2 - t_1) = m(v_2 - v_1) \qquad [\text{SI}] \qquad 72.55(a)$$

$$F(t_2 - t_1) = \frac{m(v_2 - v_1)}{q_c} \qquad [\text{U.S.}] \qquad 72.55(b)$$

$$T(t_2 - t_1) = I(\omega_2 - \omega_1) \qquad [\text{SI}] \qquad 72.56(a)$$

$$T(t_2 - t_1) = \frac{I(\omega_2 - \omega_1)}{g_c} \qquad [\text{U.S.}] \qquad 72.56(b)$$

23. IMPULSE-MOMENTUM PRINCIPLE IN OPEN SYSTEMS

$$F = \frac{m\Delta v}{\Delta t} = \dot{m}\Delta v \qquad [\text{SI}] \qquad 72.57(a)$$

$$F = \frac{m\Delta v}{g_c \Delta t} = \frac{\dot{m}\Delta v}{g_c} \qquad [\text{U.S.}] \qquad 72.57(b)$$

24. IMPACTS

$$m_1 v_1 + m_2 v_2 = m_1 v_1' + m_2 v_2' \qquad 72.58$$

$$m_1 v_1^2 + m_2 v_2^2 = m_1 v_1'^2 + m_2 v_2'^2 \big|_{\text{elastic impact}} \qquad 72.59$$

Figure 72.18 *Direct Central Impact*

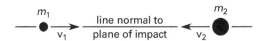

25. COEFFICIENT OF RESTITUTION

$$e = \frac{\text{relative separation velocity}}{\text{relative approach velocity}} = \frac{v_1' - v_2'}{v_2 - v_1} \qquad 72.60$$

CERM Chapter 73
Roads and Highways: Capacity Analysis

8. VOLUME PARAMETERS

$$K = \frac{\text{DHV}}{\text{AADT}} \qquad 73.1$$

$$\text{DDHV} = D(\text{DHV}) = DK(\text{AADT}) \qquad 73.2$$

$$\text{volume-capacity ratio}_i = (v/c)_i \qquad 73.3$$

$$v_{m,i} = c(v/c)_{m,i} \qquad 73.4$$

$$\text{PHF} = \frac{\text{actual hourly volume}_{\text{vph}}}{\text{peak rate of flow}_{\text{vph}}} = \frac{V_{\text{vph}}}{v_p}$$

$$= \frac{V_{\text{vph}}}{4\,V_{15\,\text{min,peak}}} \qquad 73.5$$

$$V_i = v_{m,i}N(\text{adjustment factors}) \qquad 73.6$$

9. TRIP GENERATION

$$\text{no. of trips} = a + b\left(\begin{array}{c}\text{calling population}\\\text{parameter}\end{array}\right) \qquad 73.7$$

$$\log(\text{no. of trips}) = A + B\log\left(\begin{array}{c}\text{calling population}\\\text{parameter}\end{array}\right)$$
$$73.8(a)$$

$$\text{no. of trips} = \frac{C}{\left(\begin{array}{c}\text{calling population}\\\text{parameter}\end{array}\right)^D} \qquad 73.8(b)$$

10. SPEED, FLOW, AND DENSITY RELATIONSHIPS

$$S = S_f\left(1 - \frac{D}{D_j}\right) \qquad 73.9$$

$$v = SD = \frac{3600\,\frac{\text{sec}}{\text{hr}}}{\text{headway}_{\text{sec/veh}}} \qquad 73.10$$

$$\text{spacing}_{\text{m/veh}} = \frac{1000\,\frac{\text{m}}{\text{km}}}{D_{\text{vpk/lane}}} \qquad [\text{SI}] \quad 73.11(a)$$

$$\text{spacing}_{\text{ft/veh}} = \frac{5280\,\frac{\text{ft}}{\text{mi}}}{D_{\text{vpm/lane}}} \qquad [\text{U.S.}] \quad 73.11(b)$$

$$\text{headway}_{\text{s/veh}} = \frac{\text{spacing}_{\text{m/veh}}}{\text{space mean speed}_{\text{m/s}}} \qquad [\text{SI}] \quad 73.12(a)$$

$$\text{headway}_{\text{sec/veh}} = \frac{\text{spacing}_{\text{ft/veh}}}{\text{space mean speed}_{\text{ft/sec}}} \qquad [\text{U.S.}] \quad 73.12(b)$$

$$v_{\text{vph}} = \frac{3600\,\frac{\text{sec}}{\text{hr}}}{\text{headway}_{\text{sec/veh}}} \qquad 73.13$$

13. FREEWAYS

$$D = \frac{v_p}{S} \qquad 73.14$$

$$V = v_p(\text{PHF})Nf_{\text{HV}}f_p \qquad 73.15$$

$$f_{\text{HV}} = \frac{1}{1 + P_T(E_T - 1) + P_R(E_R - 1)} \qquad 73.16$$

$$\text{FFS} = 75.4\,\frac{\text{mi}}{\text{hr}} - f_{\text{LW}} - f_{\text{LC}} - \left(3.22\,\frac{\text{mi}}{\text{hr}}\right)\text{TRD}^{0.84}$$
$$73.17$$

14. MULTILANE HIGHWAYS

$$\text{FFS} = \text{BFFS} - f_M - f_{\text{LW}} - f_{\text{LC}} - f_A \qquad 73.18$$

$$v_p = D_m(\text{FFS}) = \frac{V}{N(\text{PHF})f_{\text{HV}}f_p} \qquad 73.19$$

15. TWO-LANE HIGHWAYS

$$\text{FFS} = S_{\text{FM}} + 0.00776\,\frac{v_f}{f_{\text{HV,ATS}}} \qquad [\text{field measurements}]$$
$$73.20$$

$$\text{FFS} = \text{BFFS} - f_{\text{LS}} - f_A \qquad [\text{estimated BFFS}] \qquad 73.21$$

$$v_{i,\text{ATS}} = \frac{V_i}{(\text{PHF})f_{g,\text{ATS}}f_{\text{HV,ATS}}} \qquad [\text{ATS}] \qquad 73.22$$

$$v_{i,\text{PTSF}} = \frac{V_i}{(\text{PHF})f_{g,\text{PTSF}}f_{\text{HV,PTSF}}} \qquad [\text{PTSF}] \qquad 73.23$$

$$\text{ATS}_d = \text{FFS} - 0.0076(v_{d,s} + v_{o,s}) - f_{\text{np},s} \qquad 73.24$$

$$\text{BPTSF}_d = (1 - e^{av_d^b}) \times 100\% \qquad 73.25$$

$$\text{PTSF}_d = \text{BPTSF}_d + f_{\text{np,PTSF}}\left(\frac{v_{d,\text{PTSF}}}{v_{d,\text{PTSF}} + v_{o,\text{PTSF}}}\right) \qquad 73.26$$

$$\text{PFFS} = \frac{\text{ATS}_d}{\text{FFS}} \qquad 73.27$$

17. SIGNALIZED INTERSECTIONS

$$c_i = N s_i \left(\frac{g_i}{C}\right) \qquad 73.28$$

$$s_{\mathrm{vphgpl}} = \frac{3600 \, \dfrac{\sec}{\mathrm{hr}}}{\text{saturation headway}_{\sec/\mathrm{veh}}} \qquad 73.29$$

$$X_i = \left(\frac{v}{c}\right)_i = \frac{v_i}{N s_i \left(\dfrac{g_i}{C}\right)} = \frac{v_i C}{N s_i g_i} \qquad 73.30$$

$$X_c = \left(\frac{C}{C-L}\right) \sum_{i \in ci} y_{c,i} \qquad 73.31$$

$$L = \sum_{i \in ci} l_{t,i} \qquad 73.32$$

$$s = s_o N f_w f_{\mathrm{HV}} f_g f_p f_{\mathrm{bb}} f_a f_{\mathrm{LU}} f_{\mathrm{RT}} f_{\mathrm{LT}} f_{\mathrm{Lpb}} f_{\mathrm{Rpb}} \qquad 73.33$$

$$R_p = \frac{P_{\mathrm{green}}}{\dfrac{g}{C}} \qquad 73.34$$

18. CYCLE LENGTH: WEBSTER'S EQUATION

$$C = \frac{1.5L + 5}{1 - \sum_{\substack{\text{critical phases}}} Y_i} = \frac{1.5L + 5}{1 - \sum_{\substack{\text{critical phases}}} \left(\dfrac{v}{c}\right)_i} \qquad 73.35$$

19. CYCLE LENGTH: GREENSHIELDS METHOD

$$t_{\mathrm{phase}} = 3.8 + 2.1n \qquad 73.36$$

23. TRAFFIC-ACTIVATED TIMING

$$\begin{aligned} &\text{no. of cars in initial period} \\ &= \frac{\text{distance between line and detector}}{\text{car length}} \end{aligned} \qquad 73.37$$

24. PEDESTRIANS AND WALKWAYS

$$v = \frac{v_{p,15}}{15 W_E} \quad [HCM \text{ Eq. 23-3}] \qquad 73.38$$

$$A_p = \frac{S_p}{v_p} \quad [HCM \text{ Eq. 23-4}] \qquad 73.39$$

32. QUEUING MODELS

$$L = \lambda W \qquad 73.40$$

$$L_q = \lambda W_q \qquad 73.41$$

$$W = W_q + \frac{1}{\mu} \qquad 73.42$$

33. M/M/1 SINGLE-SERVER MODEL

$$f(t) = \mu e^{-\mu t} \qquad 73.43$$

$$P\{t > h\} = e^{-\mu h} \qquad 73.44$$

$$p\{x\} = \frac{e^{-\lambda} \lambda^x}{x!} \qquad 73.45$$

$$p\{0\} = 1 - \rho \qquad 73.46$$

$$p\{n\} = p\{0\} \rho^n \qquad 73.47$$

$$W = \frac{1}{\mu - \lambda} = W_q + \frac{1}{\mu} = \frac{L}{\lambda} \qquad 73.48$$

$$W_q = \frac{\rho}{\mu - \lambda} = \frac{L_q}{\lambda} \qquad 73.49$$

$$L = \frac{\lambda}{\mu - \lambda} = L_q + \rho \qquad 73.50$$

$$L_q = \frac{\rho \lambda}{\mu - \lambda} \qquad 73.51$$

34. M/M/s MULTI-SERVER MODEL

$$W = W_q + \frac{1}{\mu} \qquad 73.52$$

$$W_q = \frac{L_q}{\lambda} \qquad 73.53$$

$$L_q = \frac{p\{0\} \rho \left(\dfrac{\lambda}{\mu}\right)^s}{s! (1 - \rho)^2} \qquad 73.54$$

$$L = L_q + \frac{\lambda}{\mu} \qquad 73.55$$

$$p\{0\} = \frac{1}{\dfrac{\left(\dfrac{\lambda}{\mu}\right)^s}{s! \left(1 - \dfrac{\lambda}{s\mu}\right)} + \displaystyle\sum_{j=0}^{s-1} \dfrac{\left(\dfrac{\lambda}{\mu}\right)^j}{j!}} \qquad 73.56$$

$$p\{n\} = \frac{p\{0\} \left(\dfrac{\lambda}{\mu}\right)^n}{n!} \quad [n \le s] \qquad 73.57$$

$$p\{n\} = \frac{p\{0\} \left(\dfrac{\lambda}{\mu}\right)^n}{s! s^{n-s}} \quad [n > s] \qquad 73.58$$

37. TEMPORARY TRAFFIC CONTROL ZONES

$$L_m = \frac{W_m S_{km/h}^2}{155} \quad [S \le 60 \text{ km/h}] \qquad \text{[SI]} \quad 73.59(a)$$

$$L_{ft} = \frac{W_{ft} S_{mph}^2}{60} \quad [S \le 40 \text{ mph}] \qquad \text{[U.S.]} \quad 73.59(b)$$

$$L_m = \frac{W_m S_{km/h}}{1.6} \quad [S > 70 \text{ km/h}] \qquad \text{[SI]} \quad 73.60(a)$$

$$L_{ft} = W_{ft} S_{mph} \quad [S > 45 \text{ mph}] \qquad \text{[U.S.]} \quad 73.60(b)$$

CERM Chapter 75
Highway Safety

3. VEHICLE DYNAMICS

$$F_i = ma \qquad \text{[SI]} \quad 75.1(a)$$

$$F_i = \frac{ma}{g_c} = \frac{wa}{g} \qquad \text{[U.S.]} \quad 75.1(b)$$

$$F_g = w \sin \theta$$

$$\approx w \tan \theta = \frac{wG\%}{100} \qquad 75.2$$

$$F_r = f_r w \cos \phi \approx f_r w \qquad 75.3$$

$$F_D = \frac{C_D A \rho v^2}{2} \qquad \text{[SI]} \quad 75.4(a)$$

$$F_D = \frac{C_D A \rho v^2}{2g_c} \qquad \text{[U.S.]} \quad 75.4(b)$$

$$F_D = KAv^2 \approx 0.0011 A_{m^2} v_{km/h}^2 \qquad \text{[SI]} \quad 75.5(a)$$

$$F_D = KAv^2 \approx 0.0006 A_{ft^2} v_{mi/hr}^2 \qquad \text{[U.S.]} \quad 75.5(b)$$

$$v_{vehicle} = v_{t,tire} = 2\pi r_{tire} n_{tire} \qquad 75.6$$

$$n_{wheel} = \frac{n_{engine}}{R_{transmission} R_{differential}} \qquad 75.8$$

$$R_{transmission} = \frac{n_{engine}}{n_{driveshaft}} \qquad 75.9$$

$$R_{differential} = \frac{n_{driveshaft}}{n_{wheel}} \qquad 75.10$$

$$P_{kW} = \frac{T_{N \cdot m} n_{rpm}}{9549} \qquad \text{[SI]} \quad 75.11(a)$$

$$P_{hp} = \frac{T_{in-lbf} n_{rpm}}{63,025} \qquad \text{[U.S.]} \quad 75.11(b)$$

$$F_{tractive} = \frac{\eta_m T R_{transmission} R_{differential}}{r_{tire}} \qquad 75.12$$

$$v = \frac{2\pi r_{tire} n_{rev/sec}(1-i)}{R_{transmission} R_{differential}} \qquad 75.13$$

$$\dot{m}_{fuel,kg/h} = P_{brake,kW}(BSFC_{kg/kW \cdot h}) \qquad \text{[SI]} \quad 75.14(a)$$

$$\dot{m}_{fuel,lbm/hr} = P_{brake,hp}(BSFC_{lbm/hp-hr}) \qquad \text{[U.S.]} \quad 75.14(b)$$

$$\dot{Q}_{fuel,L/h} = \frac{\dot{m}_{fuel,kg/h}\left(1000 \frac{L}{m^3}\right)}{\rho_{kg/m^3}} \qquad \text{[SI]} \quad 75.15(a)$$

$$\dot{Q}_{fuel,gal/hr} = \frac{\dot{m}_{fuel,lbm/hr}\left(7.48 \frac{gal}{ft^3}\right)}{\rho_{lbm/ft^3}} \qquad \text{[U.S.]} \quad 75.15(b)$$

$$s'_{fuel,km/L} = \frac{v_{km/h}}{\dot{Q}_{fuel,L/h}} \qquad \text{[SI]} \quad 75.16(a)$$

$$s'_{fuel,mi/gal} = \frac{v_{mi/hr}}{\dot{Q}_{fuel,gal/hr}} \qquad \text{[U.S.]} \quad 75.16(b)$$

4. DYNAMICS OF STEEL-WHEELED RAILROAD ROLLING STOCK

$$DBP = TFDA - LR$$

$$= CR + AF \qquad 75.17$$

$$F_{tractive,N} = \frac{175 P_{kW,rated} \eta_{drive}}{v_{km/h}} \qquad \text{[SI]} \quad 75.18(a)$$

$$F_{tractive,lbf} = \frac{375 P_{hp,rated} \eta_{drive}}{v_{mph}} \qquad \text{[U.S.]} \quad 75.18(b)$$

$$R_{lbf/ton} = 1.3 + \frac{29}{w_{tons/axle}} + 0.045v + \frac{0.0005 A_{ft^2} v_{mph}^2}{w_{tons/axle} n} \qquad 75.19(a)$$

$$R_{lbf/car} = 1.3 W_{tons/car} + 29n + 0.45 W_{tons/car} v + 0.0005 A_{ft^2} v_{mph}^2 \qquad 75.19(b)$$

$$R_{lbf/ton} = 0.6 + \frac{20}{w_{tons}} + 0.01 v_{mph} + \frac{K v_{mph}^2}{w_{tons} n} \qquad \text{[U.S. only]} \quad 75.20$$

5. COEFFICIENT OF FRICTION

$$f = \frac{F_f}{N} \qquad 75.21$$

7. STOPPING DISTANCE

$$s_{stopping} = v t_p + s_b \qquad 75.22$$

8. BRAKING AND DECELERATION RATE

$$a = fg = f\left(9.81\,\frac{\text{m}}{\text{s}^2}\right) \qquad \text{[SI]} \quad 75.23(a)$$

$$a = fg = f\left(32.2\,\frac{\text{ft}}{\text{sec}^2}\right) \qquad \text{[U.S.]} \quad 75.23(b)$$

9. BRAKING AND SKIDDING DISTANCE

$$s_b = \frac{\text{v}_1^2 - \text{v}_2^2}{2g(f\cos\theta + \sin\theta)} \qquad 75.24$$

$$s_b = \frac{\text{v}_1^2 - \text{v}_2^2}{2g(f+G)}$$

$$\approx \frac{\text{v}_{1,\text{km/h}}^2 - \text{v}_{2,\text{km/h}}^2}{254(f+G)} \qquad \text{[SI]} \quad 75.25(a)$$

$$s_b = \frac{\text{v}_1^2 - \text{v}_2^2}{2g(f+G)}$$

$$\approx \frac{\text{v}_{1,\text{mph}}^2 - \text{v}_{2,\text{mph}}^2}{30(f+G)} \qquad \text{[U.S.]} \quad 75.25(b)$$

16. ANALYSIS OF ACCIDENT DATA

$$R_{\text{RMEV}} = \frac{(\text{no. of accidents})(10^6)}{(\text{ADT})(\text{no. of years})\left(365\,\frac{\text{days}}{\text{yr}}\right)} \qquad 75.26$$

$$R_{\text{HMVM}} = \frac{(\text{no. of accidents})(10^8)}{(\text{ADT})(\text{no. of years})\left(365\,\frac{\text{days}}{\text{yr}}\right)L_{\text{mi}}} \qquad 75.27$$

19. MODELING VEHICLE ACCIDENTS

$$\frac{m\text{v}^2}{2} = \overline{F}s \qquad \text{[SI]} \quad 75.31(a)$$

$$\frac{m\text{v}^2}{2g_c} = \overline{F}s \qquad \text{[U.S.]} \quad 75.31(b)$$

23. HSM CRASH ESTIMATION METHODS

$$N = \frac{\text{no. of crashes}}{\text{period in years}} \qquad \text{[HSM Eq. 3-1]} \quad 75.32$$

$$R = \frac{\text{average crash frequency in a period}}{\text{exposure in same period}} \qquad \text{[HSM Eq. 3-2]} \quad 75.33$$

$$N_{\text{SPF,rural,crashes/yr}} = \text{AADT}_{\text{veh/day}} \times L_{\text{mi}} \times 365 \times 10^{-6} \times e^{-0.312}$$
$$\text{[HSM Eq. 3-4]} \quad 75.34$$

$$C = \frac{\sum N_{\text{observed}}}{\sum N_{\text{predicted}}} \qquad 75.35$$

$$N_{\text{predicted}} = N_{\text{SPF}} \times C \times (\text{CMF}_1 \times \text{CMF}_2 \times \cdots \times \text{CMF}_n)$$
$$\text{[HSM Eq. 3-3]} \quad 75.36$$

24. HSM CRASH MODIFICATION FACTORS

$$\text{CMF} = \frac{\text{expected average crash frequency with modified site}}{\text{expected average crash frequency with base site}} \qquad \text{[HSM Eq. 3-5]} \quad 75.37$$

$$\text{CMF}_n = (\text{CMF}_1)^n \quad \text{[HSM Eq. 3-7]} \qquad 75.38$$

percent reduction in crash frequency
$$= (1.00 - \text{CMF}) \times 100\% \quad \text{[HSM Eq. 3-6]}$$
$$75.39$$

25. HSM NETWORK SCREENING

$$\text{excess} = (N_{\text{expected }n(\text{PDO})} - N_{\text{predicted }n(\text{PDO})})$$
$$+ (N_{\text{expected }n(\text{FI})} - N_{\text{predicted }n(\text{FI})})$$
$$\text{[HSM Eq. 4-45]} \quad 75.40$$

27. HSM ECONOMIC APPRAISAL

$$\frac{B}{C} = \frac{P_{\text{benefits}} - P_{\text{disbenefits}}}{P_{\text{cost}}} \quad \text{[HSM Eq. 7-4]} \qquad 75.41$$

$$\text{CEI} = \frac{P_{\text{cost}}}{N_{\text{predicted}} - N_{\text{observed}}} \quad \text{[HSM Eq. 7-5]} \qquad 75.42$$

CERM Chapter 76
Flexible Pavement Design

9. WEIGHT-VOLUME RELATIONSHIPS

$$P_b = (\text{aggregate surface area}) \times (\text{asphalt thickness}) \times (\text{specific weight asphalt}) \times 100\% \qquad 76.1$$

10. PLACEMENT AND PAVING EQUIPMENT

$$L_{\text{m/tonne}} = \frac{1000\,\frac{\text{kg}}{\text{tonne}}}{w_{\text{m}}t_{\text{m}}\rho_{\text{compacted,kg/m}^3}} \qquad \text{[SI]} \quad 76.2(a)$$

$$L_{\text{ft/ton}} = \frac{2000\,\frac{\text{lbf}}{\text{ton}}}{w_{\text{ft}}t_{\text{ft}}\gamma_{\text{compacted,lbf/ft}^3}} \qquad \text{[U.S.]} \quad 76.2(b)$$

$$L_{m/tonne} = \frac{1000 \; \frac{kg}{tonne}}{r_{kg/m^2} \, w_m} \qquad \text{[SI]} \quad 76.3(a)$$

$$L_{ft/ton} = \frac{18{,}000 \; \frac{lbf\text{-}ft^2}{ton\text{-}yd^2}}{r_{lbf/yd^2} \, w_{ft}} \qquad \text{[U.S.]} \quad 76.3(b)$$

$$v_{m/min} = \frac{R_{p,tonne/h} \, L_{m/tonne}}{60 \; \frac{min}{h}}$$

$$= \frac{R_{p,tonne/h} \left(1000 \; \frac{kg}{tonne}\right)\left(1000 \; \frac{mm}{m}\right)}{w_m \, t_{mm} \, \rho_{kg/m^3} \left(60 \; \frac{min}{h}\right)} \qquad \text{[SI]} \quad 76.4(a)$$

$$v_{ft/min} = \frac{R_{p,ton/hr} \, L_{ft/ton}}{60 \; \frac{min}{hr}}$$

$$= \frac{R_{p,ton/hr} \left(2000 \; \frac{lbf}{ton}\right)\left(12 \; \frac{in}{ft}\right)}{w_{ft} \, t_{in} \, \gamma_{lbf/ft^3} \left(60 \; \frac{min}{hr}\right)} \qquad \text{[U.S.]} \quad 76.4(b)$$

12. CHARACTERISTICS OF ASPHALT CONCRETE

$$G_{sa} = \frac{m}{V_{aggregate}\rho_{water}} \qquad \text{[SI]} \quad 76.5(a)$$

$$G_{sa} = \frac{W}{V_{aggregate}\gamma_{water}} \qquad \text{[U.S]} \quad 76.5(b)$$

$$G_{sb} = \frac{P_1 + P_2 + \cdots + P_n}{\dfrac{P_1}{G_1} + \dfrac{P_2}{G_2} + \cdots + \dfrac{P_n}{G_n}} \qquad 76.6$$

$$G_{sa} = \frac{A}{A - C} \qquad 76.7$$

$$G_{sb} = \frac{A}{B - C} \qquad 76.8$$

$$\text{absorption} = \frac{B - A}{A} \times 100\% \qquad 76.9$$

$$G_{sa} = \frac{A}{B + A - C} \qquad 76.10$$

$$G_{sb} = \frac{A}{B + S - C} \qquad 76.11$$

$$\text{absorption} = \frac{S - A}{A} \times 100\% \qquad 76.12$$

$$G_{se} = \frac{100\% - P_b}{\dfrac{100\%}{G_{mm}} - \dfrac{P_b}{G_b}} \qquad 76.13$$

$$G_{mm} = \frac{100\%}{\dfrac{P_s}{G_{se}} + \dfrac{P_b}{G_b}} \qquad 76.14$$

$$G_{mm} = \frac{A}{A + D - E} \qquad 76.15$$

$$P_{ba} = \frac{G_b(G_{se} - G_{sb})}{G_{sb} \, G_{se}} \times 100\% \qquad 76.16$$

$$P_{be} = P_b - \frac{P_{ba} P_s}{100\%} \qquad 76.17$$

$$VMA = 100\% - \frac{G_{mb} P_s}{G_{sb}} \qquad 76.18$$

$$P_a = VTM = \frac{G_{mm} - G_{mb}}{G_{mm}} \times 100\% \qquad 76.19$$

$$VFA = \frac{VMA - VTM}{VMA} \times 100\% \qquad 76.20$$

15. MARSHALL MIX TEST PROCEDURE

Figure 76.4 *Typical Marshall Mix Design Test Results*

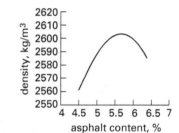

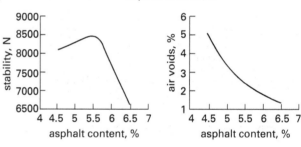

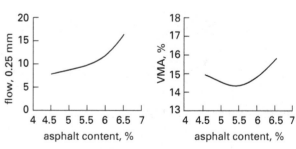

16. HVEEM MIX DESIGN

$$S = \frac{22.2}{\dfrac{p_h D}{p_v - p_h} - 0.222} \qquad 76.21$$

$$C = \frac{L}{W}(0.20H + 0.044H^2) \quad \text{[U.S. only]} \qquad 76.22$$

19. TRAFFIC

$$ESALs = (\text{no. of axles})(LEF) \qquad 76.23$$

20. TRUCK FACTORS

$$TF = \frac{ESALs}{\text{no. of trucks}} \qquad 76.24$$

21. DESIGN TRAFFIC

$$ESAL_{20} = (ESAL_{\text{first year}})(GF) \qquad 76.25$$

$$w_{18} = D_D D_L \hat{w}_{18} \qquad 76.26$$

26. PAVEMENT STRUCTURAL NUMBER

$$SN = D_1 a_1 + D_2 a_2 m_2 + D_3 a_3 m_3 \qquad 76.33$$

$$a_2 = 0.249(\log_{10} E_{BS}) - 0.977$$

$$\text{[U.S. only; AASHTO } GDPS \text{ Sec. 2.3.5]} \qquad 76.34$$

$$a_3 = 0.227(\log_{10} E_{SB}) - 0.839$$

$$\text{[U.S. only; AASHTO } GDPS \text{ Sec. 2.3.5]} \qquad 76.35$$

CERM Chapter 77
Rigid Pavement Design

3. AASHTO METHOD OF RIGID PAVEMENT DESIGN

$$\log_{10} w_{18} = z_R S_o + 7.35 \log_{10}(D+1) - 0.06$$

$$+ \frac{\log_{10}\left(\dfrac{\Delta PSI}{4.5 - 1.5}\right)}{1 + \dfrac{1.624 \times 10^7}{(D+1)^{8.46}}}$$

$$+ (4.22 - 0.32 p_t)$$

$$\times \log_{10}\left(\frac{S_c' C_d (D^{0.75} - 1.132)}{215.63 J \left(D^{0.75} - \dfrac{18.42}{\left(\dfrac{E_c}{k}\right)^{0.25}}\right)}\right)$$

$$77.1$$

4. LAYER MATERIAL STRENGTHS

$$k_{\text{lbf/in}^3} = \frac{M_{R,\text{lbf/in}^2}}{19.4} \qquad 77.2$$

$$M_R = K\sqrt{f_c'} \qquad 77.3$$

$$E_c = 57{,}000\sqrt{f_c'} \quad [E_c \text{ and } f_c' \text{ in lbf/in}^2] \qquad 77.4$$

8. STEEL REINFORCING

$$P_s = \frac{L_{\text{ft}} F}{2 f_{s,\text{lbf/in}^2}} \times 100\% \qquad 77.5$$

$$P_t = \frac{A_s}{YD} \times 100\% \qquad 77.6$$

CERM Chapter 78
Plane Surveying

1. ERROR ANALYSIS: MEASUREMENTS OF EQUAL WEIGHT

$$x_p = \frac{x_1 + x_2 + \cdots + x_k}{k} \qquad 78.1$$

$$E_{\text{mean}} = \frac{0.6745 s}{\sqrt{k}}$$

$$= \frac{E_{\text{total}, k \text{ measurements}}}{\sqrt{k}} \qquad 78.2$$

$$LCL_{50\%} = x_p - E_{\text{mean}} \qquad 78.3$$

$$UCL_{50\%} = x_p + E_{\text{mean}} \qquad 78.4$$

2. ERROR ANALYSIS: MEASUREMENTS OF UNEQUAL WEIGHT

$$E_{p,\text{weighted}} = 0.6745\sqrt{\frac{\sum\left(w_i(\overline{x} - x_i)^2\right)}{(k-1)\sum w_i}} \qquad 78.5$$

3. ERRORS IN COMPUTED QUANTITIES

$$E_{\text{sum}} = \sqrt{E_1^2 + E_2^2 + E_3^2 + \cdots} \qquad 78.6$$

$$E_{\text{product}} = \sqrt{x_1^2 E_2^2 + x_2^2 E_1^2} \qquad 78.7$$

13. DISTANCE MEASUREMENT: TAPING

$$C_T = L\alpha(T - T_s) \quad \text{[temperature]} \qquad 78.8$$

$$C_P = \frac{(P - P_s)L}{AE} \quad \text{[tension]} \qquad 78.9$$

$$C_s = \pm\frac{W^2 L^3}{24 P^2} \quad \text{[sag]} \qquad 78.10$$

14. DISTANCE MEASUREMENT: TACHEOMETRY

$$x = K(R_2 - R_1) + C \qquad 78.11$$

$$x = K(R_2 - R_1)\cos^2\theta + C\cos\theta \qquad 78.12$$

$$y = \tfrac{1}{2}K(R_2 - R_1)\sin 2\theta + C\sin\theta \qquad 78.13$$

Figure 78.1 Horizontal Stadia Measurement

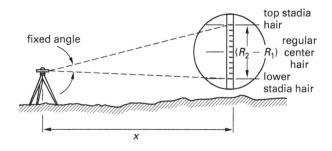

18. ELEVATION MEASUREMENT

$$h_c = \left(7.84 \times 10^{-8}\ \frac{1}{\text{m}}\right) x_{\text{m}}^2 \qquad \text{[SI]} \quad 78.14(a)$$

$$h_c = \left(2.39 \times 10^{-8}\ \frac{1}{\text{ft}}\right) x_{\text{ft}}^2 \qquad \text{[U.S.]} \quad 78.14(b)$$

$$h_r = \left(10.8 \times 10^{-9}\ \frac{1}{\text{m}}\right) x_{\text{m}}^2 \qquad \text{[SI]} \quad 78.15(a)$$

$$h_r = \left(3.3 \times 10^{-9}\ \frac{1}{\text{ft}}\right) x_{\text{ft}}^2 \qquad \text{[U.S.]} \quad 78.15(b)$$

$$h_{rc,\text{m}} = |h_r - h_c| = \left(6.75 \times 10^{-8}\ \frac{1}{\text{m}}\right) x_{\text{m}}^2 \quad \text{[SI]} \quad 78.16(a)$$

$$h_{rc,\text{ft}} = |h_r - h_c| = \left(2.06 \times 10^{-8}\ \frac{1}{\text{ft}}\right) x_{\text{ft}}^2 \quad \text{[U.S.]} \quad 78.16(b)$$

$$h_a = R_{\text{observed}} - h_{rc} \qquad 78.17$$

19. ELEVATION MEASUREMENT: DIRECT LEVELING

$$y_{\text{A-L}} = R_{\text{A}} - h_{rc,\text{A-L}} - \text{HI} \qquad 78.18$$

$$y_{\text{L-B}} = \text{HI} + h_{rc,\text{L-B}} - R_{\text{B}} \qquad 78.19$$

$$y_{\text{A-B}} = y_{\text{A-L}} + y_{\text{L-B}}$$

$$= R_{\text{A}} - R_{\text{B}} + h_{rc,\text{L-B}} - h_{rc,\text{A-L}} \qquad 78.20$$

$$y_{\text{A-B}} = R_{\text{A}} - R_{\text{B}} \qquad 78.21$$

Figure 78.4 Direct Leveling

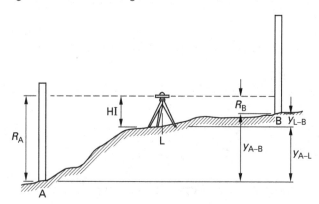

21. ELEVATION MEASUREMENT: INDIRECT LEVELING

$$y_{\text{A-B}} = \text{AC}\tan\alpha + 2.1 \times 10^{-8}(\text{AC})^2 \qquad \text{[U.S.]} \quad 78.22$$

$$y_{\text{A-B}} = \text{AC}\tan\beta - 2.1 \times 10^{-8}(\text{AC})^2 \qquad \text{[U.S.]} \quad 78.23$$

$$y_{\text{A-B}} = \tfrac{1}{2}\text{AC}(\tan\alpha + \tan\beta) \qquad 78.24$$

Figure 78.5 Indirect Leveling

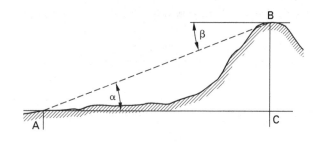

23. DIRECTION SPECIFICATION

$$\tan B = \frac{x_2 - x_1}{y_2 - y_1} \qquad 78.25$$

$$D = \sqrt{(y_2 - y_1)^2 + (x_2 - x_1)^2} \qquad 78.26$$

$$\tan A_{\text{N}} = \frac{x_2 - x_1}{y_2 - y_1} \qquad 78.27$$

$$\tan A_{\text{S}} = \frac{x_1 - x_2}{y_1 - y_2} \qquad 78.28$$

Figure 78.6 Calculation of Bearing Angle

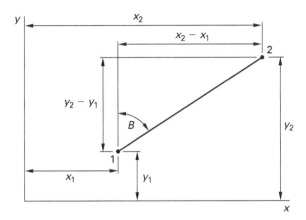

25. TRAVERSES

1. The sum of the deflection angles is 360°.
2. The sum of the interior angles of a polygon with n sides is $(n-2)(180°)$.

28. BALANCING CLOSED TRAVERSE DISTANCES

$$L = \sqrt{\begin{array}{c}(\text{closure in departure})^2 \\ + (\text{closure in latitude})^2\end{array}} \qquad 78.31$$

$$\frac{\text{leg departure correction}}{\text{closure in departure}} = \frac{-\text{leg length}}{\text{total traverse length}} \qquad 78.32$$

$$\frac{\text{leg latitude correction}}{\text{closure in latitude}} = \frac{-\text{leg length}}{\text{total traverse length}} \qquad 78.33$$

$$\frac{\text{leg departure correction}}{\text{closure in departure}}$$
$$= -\left(\frac{\text{leg departure}}{\text{sum of departure absolute values}}\right) \qquad 78.34$$

$$\frac{\text{leg latitude correction}}{\text{closure in latitude}}$$
$$= -\left(\frac{\text{leg latitude}}{\text{sum of latitude absolute values}}\right) \qquad 78.35$$

30. TRAVERSE AREA: METHOD OF COORDINATES

$$A = \frac{1}{2}\left|\sum_{i=1}^{n} y_i(x_{i-1} - x_{i+1})\right| \qquad 78.36$$

$$\frac{x_1}{y_1} \diagdown \frac{x_2}{y_2} \diagdown \frac{x_3}{y_3} \diagdown \frac{x_4}{y_4} \diagdown \frac{x_1}{y_1} \cdots$$

$$A = \frac{1}{2}\left|\begin{array}{c}\sum \text{ of full line products} \\ -\sum \text{ of broken line products}\end{array}\right| \qquad 78.37$$

31. TRAVERSE AREA: DOUBLE MERIDIAN DISTANCE

$$\mathrm{DMD}_{\text{leg }i} = \mathrm{DMD}_{\text{leg }i-1} + \text{departure}_{\text{leg }i-1}$$
$$+ \text{departure}_{\text{leg }i} \qquad 78.38$$

$$A = \frac{1}{2}\left|\sum(\text{latitude}_{\text{leg }i} \times \mathrm{DMD}_{\text{leg }i})\right| \qquad 78.39$$

32. AREAS BOUNDED BY IRREGULAR BOUNDARIES

$$A = d\left(\frac{h_1 + h_n}{2} + \sum_{i=2}^{n-1} h_i\right) \qquad 78.40$$

$$A = \frac{d}{3}\left(h_1 + h_n + 2\sum_{\text{odd}} h_i + 4\sum_{\text{even}} h_i\right)$$
$$= \frac{d}{3}(h_1 + 4h_2 + 2h_3 + 4h_4 + \ldots + h_n) \qquad 78.41$$

33. PHOTOGRAMMETRY

$$S = \frac{f}{H_{\text{AGL}}} = \frac{f}{H_{\text{MSL}} - E_{\text{MSL}}}$$
$$= \frac{\text{length in photograph}}{\text{true length}} \qquad 78.42$$

34. DETERMINING OBJECT HEIGHT PHOTOGRAMMETRICALLY

$$h_{\text{AGL}} = l\tan\alpha \qquad 78.43$$

$$h_{\text{AGL}} = \frac{dH_{\text{AGL}}}{r} \qquad 78.44$$

$$h_{\text{AGL}} = \frac{H_{\text{AGL}}(\mathrm{dP})}{P + \mathrm{dP}} \qquad 78.45$$

CERM Chapter 79
Horizontal, Compound, Vertical, and Spiral Curves

1. HORIZONTAL CURVES

(See Fig. 79.1.)

$$R = \frac{5729.578 \, \frac{\text{ft}}{\text{deg}}}{D} \qquad \text{[U.S.; arc definition]} \quad 79.1$$

$$R = \frac{50 \text{ ft}}{\sin\frac{D}{2}} \qquad \text{[U.S.; chord definition]} \quad 79.2$$

$$L = \frac{2\pi RI}{360°} = RI_{\text{radians}} = \frac{(100 \text{ ft})I}{D} \qquad \text{[U.S.]} \quad 79.3$$

$$T = R\tan\frac{I}{2} \qquad 79.4$$

$$E = R\left(\sec\frac{I}{2} - 1\right) = R\tan\frac{I}{2}\tan\frac{I}{4} \qquad 79.5$$

$$M = R\left(1 - \cos\frac{I}{2}\right) = \frac{C}{2}\tan\frac{I}{4} \qquad 79.6$$

$$C = 2R\sin\frac{I}{2} = 2T\cos\frac{I}{2} \qquad 79.7$$

Figure 79.3 *Circular Curve Deflection Angle*

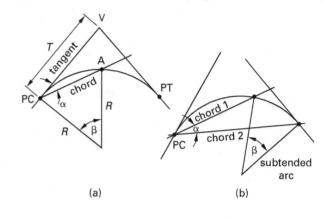

Figure 79.1 *Horizontal Curve Elements*

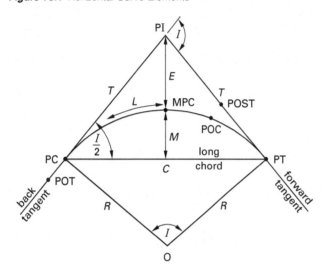

$$\alpha = \angle\text{V-PC-A} = \frac{\beta}{2} \qquad 79.13$$

$$\frac{\beta}{360°} = \frac{\text{arc length PC-A}}{2\pi R} \qquad 79.14$$

$$\frac{\beta}{I} = \frac{\text{arc length PC-A}}{L} \qquad 79.15$$

$$C_{\text{PC-A}} = 2R\sin\alpha = 2R\sin\frac{\beta}{2} \qquad 79.16$$

2. DEGREE OF CURVE

$$D = \frac{(360°)(100\text{ ft})}{2\pi R} = \frac{5729.578\text{ ft}}{R} \quad \text{[arc basis]} \qquad 79.8$$

$$\sin\frac{D}{2} = \frac{50}{R} \quad \text{[chord basis]} \qquad 79.9$$

$$L \approx \left(\frac{I}{D}\right)(100\text{ ft}) \quad [D \le 4°] \qquad 79.10$$

3. STATIONING ON A HORIZONTAL CURVE

$$\text{sta PT} = \text{sta PC} + L \qquad 79.11$$
$$\text{sta PC} = \text{sta PI} - T \qquad 79.12$$

4. CURVE LAYOUT BY DEFLECTION ANGLE

1. The deflection angle between a tangent and a chord is half of the arc's subtended angle, as shown in Fig. 79.3(a).

2. The angle between two chords is half of the arc's subtended angle, as shown in Fig. 79.3(b).

5. TANGENT OFFSET

$$y = R(1 - \cos\beta)$$
$$= R - \sqrt{R^2 - x^2} \qquad 79.17$$

$$\beta = \arcsin\frac{x}{R} = \arccos\left(\frac{R-y}{R}\right) \qquad 79.18$$

$$x = R\sin\beta = \sqrt{2Ry - y^2} \qquad 79.19$$

$$\frac{y}{W} = \frac{x^2}{L^2} \qquad 79.20$$

Figure 79.4 *Tangent Offset*

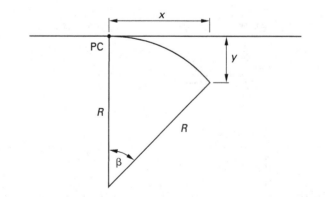

6. CURVE LAYOUT BY TANGENT OFFSET

$$\text{NP} = \text{tangent distance} = \text{NQ}\cos\alpha \qquad 79.21$$

$$\text{PQ} = \text{tangent offset} = \text{NQ}\sin\alpha \qquad 79.22$$

$$\text{NQ} = C = 2R\sin\alpha \qquad 79.23$$

$$\text{NP} = (2R\sin\alpha)\cos\alpha$$
$$= C\cos\alpha \qquad 79.24$$

$$\text{PQ} = (2R\sin\alpha)\sin\alpha$$
$$= 2R\sin^2\alpha \qquad 79.25$$

Figure 79.5 Tangent and Chord Offset Geometry

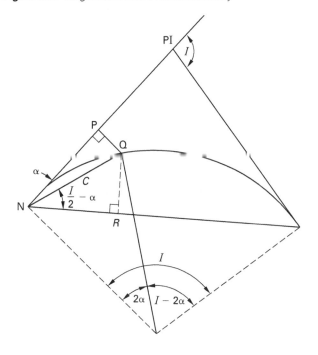

7. CURVE LAYOUT BY CHORD OFFSET

$$\text{NR} = \text{chord distance} = \text{NQ}\cos\left(\frac{I}{2}-\alpha\right)$$

$$= (2R\sin\alpha)\cos\left(\frac{I}{2}-\alpha\right)$$

$$= C\cos\left(\frac{I}{2}-\alpha\right) \qquad 79.26$$

$$\text{RQ} = \text{chord offset} = \text{NQ}\sin\left(\frac{I}{2}-\alpha\right)$$

$$= (2R\sin\alpha)\sin\left(\frac{I}{2}-\alpha\right)$$

$$= C\sin\left(\frac{I}{2}-\alpha\right) \qquad 79.27$$

8. HORIZONTAL CURVES THROUGH POINTS

$$\alpha = \arctan\frac{y}{x} \qquad 79.28$$

$$m = \sqrt{x^2+y^2} \qquad 79.29$$

$$\gamma = 90° - \frac{I}{2} - \alpha \qquad 79.30$$

$$\phi = 180° - \arcsin\left(\frac{\sin\gamma}{\cos\dfrac{I}{2}}\right) \qquad 79.31$$

$$\theta = 180° - \gamma - \phi \qquad 79.32$$

$$\frac{\sin\theta}{m} = \frac{\sin\phi\cos\dfrac{I}{2}}{R} \qquad 79.33$$

Figure 79.6 Horizontal Curve Through a Known Point

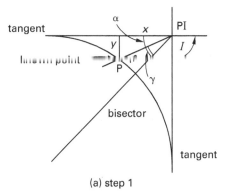

(a) step 1

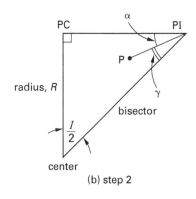

(b) step 2

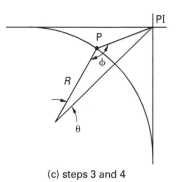

(c) steps 3 and 4

10. SUPERELEVATION

$$F_c = \frac{mv_t^2}{R} \qquad \text{[SI]} \quad 79.34(a)$$

$$F_c = \frac{mv_t^2}{g_c R} \qquad \text{[U.S.]} \quad 79.34(b)$$

$$e = \tan\phi = \frac{v^2}{gR} \quad \text{[consistent units]} \qquad 79.35$$

$$e = \tan\phi = \frac{v^2}{gR} - f_s \quad \text{[consistent units]} \qquad 79.36$$

$$e = \tan\phi = \frac{v_{km/h}^2}{127R} - f_s \qquad \text{[SI]} \quad 79.37(a)$$

$$e = \tan\phi = \frac{v_{mph}^2}{15R} - f_s \qquad \text{[U.S.]} \quad 79.37(b)$$

$$f_s = 0.16 - \frac{0.01(v_{mph} - 30)}{10} \quad \text{[< 50 mph]} \qquad 79.38$$

$$f_s = 0.14 - \frac{0.02(v_{mph} - 50)}{10} \quad \text{[50-70 mph]} \qquad 79.39$$

12. TRANSITIONS TO SUPERELEVATION

$$T_R = \frac{wp}{\text{SRR}} \qquad 79.40$$

$$L = \frac{we}{\text{SRR}} \qquad 79.41$$

13. SUPERELEVATION OF RAILROAD LINES

$$E = \frac{G_{eff}v^2}{gR} \quad \text{[railroads]} \qquad 79.42$$

14. STOPPING SIGHT DISTANCE

$$S = \left(0.278 \frac{\frac{m}{s}}{\frac{km}{h}}\right) t_p v_{km/h} + \frac{v_{km/h}^2}{254(f+G)} \qquad \text{[SI]} \quad 79.43(a)$$

$$S = \left(1.47 \frac{\frac{ft}{sec}}{\frac{mi}{hr}}\right) t_p v_{mph} + \frac{v_{mph}^2}{30(f+G)} \qquad \text{[U.S.]} \quad 79.43(b)$$

16. MINIMUM HORIZONTAL CURVE LENGTH FOR STOPPING DISTANCE

$$S = \left(\frac{R}{28.65}\right)\left(\arccos\frac{R - \text{HSO}}{R}\right) \qquad 79.44$$

$$\text{HSO} = R(1 - \cos\theta) = R\left(1 - \cos\frac{DS}{200}\right)$$

$$= R\left(1 - \cos\frac{28.65S}{R}\right) \qquad 79.45$$

17. VERTICAL CURVES

$$R = \frac{G_2 - G_1}{L} \quad \text{[may be negative]} \qquad 79.46$$

$$\text{elev}_x = \frac{Rx^2}{2} + G_1 x + \text{elev}_{BVC} \qquad 79.47$$

$$x_{\text{turning point}} = \frac{-G_1}{R} \quad \text{[in stations]} \qquad 79.48$$

$$M_{ft} = \frac{AL_{sta}}{8} \qquad 79.49$$

Figure 79.10 *Symmetrical Parabolic Vertical Curve*

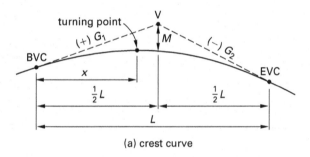

(a) crest curve

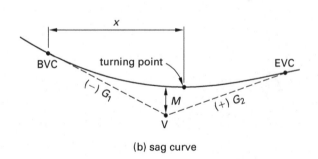

(b) sag curve

18. VERTICAL CURVES THROUGH POINTS

$$s = \sqrt{\frac{\text{elev}_E - \text{elev}_G}{\text{elev}_E - \text{elev}_F}} \qquad 79.50$$

$$L = \frac{2d(s+1)}{s-1} \qquad 79.51$$

Figure 79.11 *Vertical Curve with an Obstruction*

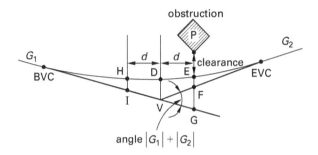

19. VERTICAL CURVE TO PASS THROUGH TURNING POINT

$$L = \frac{2(G_2 - G_1)(\text{elev}_V - \text{elev}_{TP})}{G_1 G_2}$$

$$= \frac{2(G_2 - G_1)(\text{elev}_{BVC} - \text{elev}_{TP})}{G_1^2} \qquad 79.52$$

20. MINIMUM VERTICAL CURVE LENGTH FOR SIGHT DISTANCES (CREST CURVES)

$$L = \frac{AS^2}{100(\sqrt{2h_1} + \sqrt{2h_2})^2} \quad [S < L] \qquad 79.53$$

$$L = 2S - \frac{200(\sqrt{h_1} + \sqrt{h_2})^2}{A} \quad [S > L] \qquad 79.54$$

$$L = \frac{AS^2}{800(C - 1.5)} \quad [S < L] \qquad \text{[SI]} \quad 79.55(a)$$

$$L = \frac{AS^2}{800(C - 5)} \quad [S < L] \qquad \text{[U.S.]} \quad 79.55(b)$$

$$L = 2S - \frac{800(C - 1.5)}{A} \quad [S > L] \qquad \text{[SI]} \quad 79.56(a)$$

$$L = 2S - \frac{800(C - 5)}{A} \quad [S > L] \qquad \text{[U.S.]} \quad 79.56(b)$$

Table 79.4 *AASHTO Required Lengths of Curves on Grades*[a]

	stopping sight distance[b] (crest curves)	passing sight distance[c] (crest curves)	stopping sight distance (sag curves)
SI units			
$S < L$	$L = \dfrac{AS^2}{658}$	$L = \dfrac{AS^2}{864}$	$L = \dfrac{AS^2}{120 + 3.5S}$
$S > L$	$L = 2S - \dfrac{658}{A}$	$L = 2S - \dfrac{864}{A}$	$L = 2S - \dfrac{120 + 3.5S}{A}$
U.S. units			
$S < L$	$L = \dfrac{AS^2}{2158}$	$L = \dfrac{AS^2}{2800}$	$L = \dfrac{AS^2}{400 + 3.5S}$
$S > L$	$L = 2S - \dfrac{2158}{A}$	$L = 2S - \dfrac{2800}{A}$	$L = 2S - \dfrac{400 + 3.5S}{A}$

[a] $A = |G_2 - G_1|$, absolute value of the algebraic difference in grades, in percent.
[b] The driver's eye is 3.5 ft (1080 mm) above road surface, viewing an object 2.0 ft (600 mm) high.
[c] The driver's eye is 3.5 ft (1080 mm) above road surface, viewing an object 3.5 ft (1080 mm) high.

Compiled from *A Policy on Geometric Design of Highways and Streets*, Chap. 3, copyright © 2011 by the American Association of State Highway and Transportation Officials, Washington, D.C.

21. DESIGN OF CREST CURVES USING *K*-VALUE

$$K = \frac{L}{A} = \frac{L}{|G_2 - G_1|} \quad \text{[always positive]} \qquad 79.57$$

23. MINIMUM VERTICAL CURVE LENGTH FOR COMFORT: SAG CURVES

$$L_m = \frac{A\text{v}_{\text{km/h}}^2}{395} \qquad \text{[SI]} \quad 79.58(a)$$

$$L_{ft} = \frac{A\text{v}_{\text{mph}}^2}{46.5} \qquad \text{[U.S.]} \quad 79.58(b)$$

26. SPIRAL CURVES

$$L_{s,m} \approx \frac{0.0214\text{v}_{\text{km/h}}^3}{R_m C_{\text{m/s}^3}} \qquad \text{[SI]} \quad 79.62(a)$$

$$L_{s,ft} \approx \frac{3.15\text{v}_{\text{mph}}^3}{R_{ft} C_{\text{ft/sec}^3}} \qquad \text{[U.S.]} \quad 79.62(b)$$

$$I_s = \left(\frac{L_s}{100}\right)\left(\frac{D}{2}\right) = \frac{L_s D}{200} \qquad 79.63$$

$$I = I_c + 2I_s \qquad 79.64$$

$$L = L_c + 2L_s \qquad 79.65$$

$$\alpha_s = \tan^{-1}\frac{y}{x} \approx \frac{y}{x} \approx \frac{I_s}{3} \quad [I_s < 20°] \qquad 79.66$$

$$\frac{\alpha_P}{\alpha_s} = \frac{I_P}{I_s} = \left(\frac{L_P}{L_s}\right)^2 \qquad 79.67$$

$$y = x\tan\alpha_P \approx x\alpha_{P,\text{radians}}$$

$$= \frac{xI_{s,\text{radians}}}{3}$$

$$= \left(\frac{xL_s D_{\text{degrees}}}{(200)(3)}\right)\left(\frac{\pi}{180°}\right)$$

$$= \frac{xL_s}{6R_c} \quad [I_s < 20°] \qquad 79.68$$

$$T_c = (R_c + Q)\tan\frac{I}{2} \qquad 79.69$$

$$Q = y - R_c(1 - \cos I_s)$$

$$= \frac{L_s^2}{6R_c} - R_c(1 - \cos I_s) \qquad 79.70$$

Figure 79.17 *Spiral Curve Geometry*

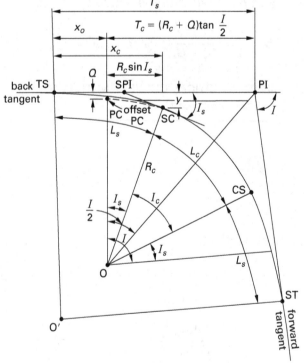

27. SPIRAL LENGTH TO PREVENT LANE ENCROACHMENT

$$L_{s,\text{min,m}} = \sqrt{24p_{\text{min,m}}R_\text{m}} \qquad [\text{SI}] \quad 79.71(a)$$

$$L_{s,\text{min,ft}} = \sqrt{24p_{\text{min,ft}}R_\text{ft}} \qquad [\text{U.S.}] \quad 79.71(b)$$

$$L_{s,\text{max,m}} = \sqrt{24p_{\text{max,m}}R_\text{m}} \qquad [\text{SI}] \quad 79.72(a)$$

$$L_{s,\text{max,ft}} = \sqrt{24p_{\text{max,ft}}R_\text{ft}} \qquad [\text{U.S.}] \quad 79.72(b)$$

28. MAXIMUM RADIUS FOR USE OF A SPIRAL

$$R_{\text{max,m}} = \frac{v^2}{a} = 0.0592v_{\text{km/h}}^2 \qquad [\text{SI}] \quad 79.73(a)$$

$$R_{\text{max,ft}} = \frac{v^2}{a} = 0.506v_{\text{mph}}^2 \qquad [\text{U.S.}] \quad 79.73(b)$$

29. DESIRABLE SPIRAL CURVE LENGTH

$$L_{s,\text{desirable,m}} = 0.556v_{\text{km/h}} \qquad [\text{SI}] \quad 79.74(a)$$

$$L_{s,\text{desirable,ft}} = 2.93v_{\text{mph}} \qquad [\text{U.S.}] \quad 79.74(b)$$

Construction

3. SWELL AND SHRINKAGE

(See Table 80.1.)

$$V_l = \left(\frac{100\% + \% \, \text{swell}}{100\%} \right) V_b = \frac{V_b}{L} \qquad \textit{80.1}$$

$$V_c = \left(\frac{100\% - \% \, \text{shrinkage}}{100\%} \right) V_b \qquad \textit{80.2}$$

12. VOLUMES OF PILES

$$V = \left(\frac{h}{3} \right) \pi r^2 \quad [\text{cone}] \qquad \textit{80.3}$$

$$V = \left(\frac{h}{6} \right) b(2a + a_1) = \tfrac{1}{6} h b \left(3a - \frac{2h}{\tan \phi} \right)$$
$$[\text{wedge}] \qquad \textit{80.4}$$

Table 80.1 Summary of Excavation Soil Factors

quantity	symbol	example value or units	formulas		
swell factor; heaped factor [bank to loose]	SF	1.20	$\dfrac{\text{LCY}}{\text{BCY}}$	$\dfrac{1}{\text{LF}}$	$\dfrac{\gamma_{\text{bank}}}{\gamma_{\text{loose}}}$
shrinkage factor; shrink factor [bank to compacted]	DF	0.15	$1 - \dfrac{\text{CCY}}{\text{BCY}}$		$1 - \dfrac{\gamma_{\text{bank}}}{\gamma_{\text{compacted}}}$
load factor [loose to bank]	LF	0.83	$\dfrac{\text{BCY}}{\text{LCY}}$	$\dfrac{1}{\text{SF}}$	$\dfrac{\gamma_{\text{loose}}}{\gamma_{\text{bank}}}$
fill factor; dipper factor; fillability	FF	0.85	$\dfrac{\text{occupied volume}}{\text{solid capacity of bucket, dipper, or truck}}$		
operator factor	OF	0.93	$\dfrac{\text{actual production}}{\text{ideal production}}$		
loose volume; loose cubic yards	LCY	120 yd^3	$\text{SF} \times \text{BCY}$	$\dfrac{\text{BCY}}{\text{LF}}$	$\dfrac{\gamma_{\text{bank}}}{\gamma_{\text{loose}}} \times \text{BCY}$
compacted volume; compacted cubic yards	CCY	85 yd^3	$(1 - \text{DF})\text{BCY}$	$\left(\dfrac{1 - \text{DF}}{\text{SF}} \right)\text{LCY}$	$\dfrac{\gamma_{\text{bank}}}{\gamma_{\text{compacted}}} \times \text{BCY}$
bank volume; bank cubic yards	BCY	100 yd^3	$\dfrac{\text{LCY}}{\text{SF}}$	$\dfrac{\text{CCY}}{1 - \text{DF}}$	
swell	SWL	20 yd^3	$\text{LCY} - \text{BCY}$ (increase)	$(\text{SF} - 1)\text{BCY}$	$\dfrac{1 - \text{LF}}{\text{LF}} \times \text{BCY}$
		20%	$\dfrac{\text{LCY} - \text{BCY}}{\text{BCY}} \times 100\%$ (increase)	$(\text{SF} - 1) \times 100\%$	$\dfrac{1 - \text{LF}}{\text{LF}} \times 100\%$
shrinkage	SHR	15 yd^3	$\text{BCY} - \text{CCY}$ (decrease)	$\text{DF} \times \text{BCY}$	
		15%	$\dfrac{\text{BCY} - \text{CCY}}{\text{BCY}} \times 100\%$ (decrease)	$\text{DF} \times 100\%$	

$$V = \left(\frac{h}{6}\right)\left(ab + (a + a_1)(b + b_1) + a_1 b_1\right)$$

$$= \left(\frac{h}{6}\right)\left(ab + 4\left(a - \frac{h}{\tan\phi_1}\right)\left(b - \frac{h}{\tan\phi_2}\right)\right.$$

$$\left. + \left(a - \frac{2h}{\tan\phi_1}\right)\left(b - \frac{2h}{\tan\phi_2}\right)\right)$$

$$\begin{bmatrix} \text{frustum of a} \\ \text{rectangular pyramid} \end{bmatrix} \qquad 80.5$$

$$a_1 = a - \frac{2h}{\tan\phi_1} \qquad 80.6$$

$$b_1 = b - \frac{2h}{\tan\phi_2} \qquad 80.7$$

Figure 80.5 *Pile Shapes*

(a) cone

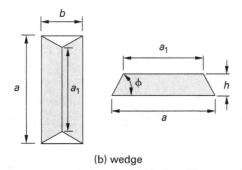

(b) wedge

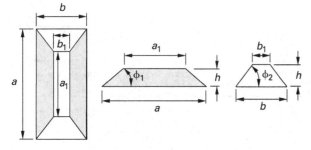

(c) frustum of a rectangular pyramid

14. AVERAGE END AREA METHOD

$$V = \frac{L(A_1 + A_2)}{2} \qquad 80.8$$

$$V_{\text{pyramid}} = \frac{LA_{\text{base}}}{3} \qquad 80.9$$

15. PRISMOIDAL FORMULA METHOD

$$V = \left(\frac{L}{6}\right)(A_1 + 4A_m + A_2) \qquad 80.10$$

17. MASS DIAGRAMS

Figure 80.9 *Balance Line Between Two Points*

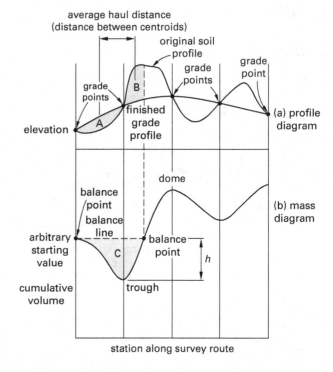

CERM Chapter 81
Construction Staking and Layout

2. STAKE MARKINGS

(See Table 81.2.)

3. ESTABLISHING SLOPE STAKE MARKINGS

$$\text{HI} = \text{elev}_{\text{ground}} + \frac{\text{instrument height}}{\text{above the ground}}$$

$$= \text{elev}_{\text{ground}} + \text{ground rod} \qquad 81.1$$

$$\text{grade rod} = \text{HI} - \text{elev}_{\text{grade}} \qquad 81.2$$

$$h = \text{grade rod} - \text{ground rod} \qquad 81.3$$

$$d = \frac{w}{2} + hs \qquad 81.4$$

Table 81.2 Common Stake Marking Abbreviations*

@	from the	HP	hinge point
$1/4$SR	quarter point of slope rounding	INL	inlet
$1/2$SR	midpoint of slope rounding	INT	intersection
ABUT	abutment	INV	invert
AHD	ahead	ISS	intermediate slope stake
BC	begin curve or back of curb	JT	joint trench
BCH	bench	L	length or left
BCR	begin curb return	L/2	midpoint
BEG	begin or beginning	L/4	quarter point
BK	back	LIP	lip
BL	baseline	L/O	line only
BM	benchmark	LOL	lay-out line
BR	bridge	LT	left
BSR	begin slope rounding	MC	middle of curve
BSW	back of sidewalk	MH	manhole
BU	build up	MP	midpoint
BVC	begin vertical curve	MSR	midpoint of slope rounding
C	cut	OG	original ground
CB	catch basin	O/S	offset
CF	curb face	PC	point of curvature (point of curve)
CGS	contour grading stake	PCC	point of compound curve
CHNL	channel	PG	pavement grade or profile grade
CL	centerline	PI	point of intersection
CONT	contour	POC	point on curve
CP	control point or catch point	POL	point on line
CR	curb return	POT	point on tangent
CS	curb stake	PP	power pole
CURB	curb	PPP	pavement plane projected
DAY	daylight	PRC	point of reverse curvature
DI	drop inlet or drainage inlet	PT	point of tangency
DIT	ditch	PVC	point of vertical curvature
DL	daylight	PVT	point of vertical tangency
D/L	daylight	QSR	quarterpoint of slope rounding
DMH	drop manhole	R	radius
DS	drainage stake	RGS	rough grade stake
E	flow line	RP	radius point or reference point
EC	end of curve	RPSS	reference point for slope stake
ECR	end of curb return	ROW	right of way
EL	elevation	RT	right
ELECT	electrical	R/W	right of way
ELEV	elevation	SD	storm drain
END	end or ending	SE	superelevation
EOM	edge of metal (EOP)	SG	subgrade
EOP	edge of pavement	SHLD	shoulder
EP	edge of pavement	SHO	shoulder
ES	edge of shoulder	SL	stationing line or string line
ESR	end slope rounding	SR	slope rounding
ETW	edge of traveled way	SS	sanitary sewer or slope stake
EVC	end of vertical curve	STA	station
EW	end wall	STR	structure
F	fill	SW	sidewalk
FC	face of curb	TBC	top back of curb
FDN	foundation	TBM	temporary benchmark
FE	fence	TC	toe of curb or top of curve
FG	finish grade	TOE	toe
FGS	final grade stake	TOP	top
FH	fire hydrant	TP	turning point
FL	flow line or flared	TW	traveled way
FLC	flow line curb	VP	vent pipe
FTG	footing	WALL	wall
G	grade	WL	water line
GRT	grate	WM	water meter
GS	gutter slope	WV	water valve
GTR	gutter	WW	wingwall
GUT	gutter		

*Compiled from various sources. Not intended to be exhaustive. Differences may exist from agency to agency.

CERM Chapter 82
Building Codes and Materials Testing

8. REQUIREMENTS BASED ON OCCUPANCY

(See Table 82.2.)

CERM Chapter 83
Construction and Job Site Safety

5. SOIL CLASSIFICATION

OSHA Soil Classifications
(OSHA 1926 Subpart P App. A)

Type A soils are cohesive soils with an unconfined compressive strength of 1.5 tons per square foot (144 kPa) or greater. Examples of type A cohesive soils are clay, silty clay, sandy clay, clay loam, and in some cases, silty clay loam and sandy clay loam. No soil is type A if it is fissured, is subject to vibration of any type, has

Table 82.2 Occupancy Groups Summary*

occupancy	description	examples
A-1	assembly with fixed seats for viewing of performances or movies	movie theaters, live performance theaters
A-2	assembly for food and drink consumption	bars, restaurants, clubs
A-3	assembly for worship, recreation, etc., not classified elsewhere	libraries, art museums, conference rooms with more than 50 occupants
A-4	assembly for viewing of indoor sports	arenas
A-5	assembly for outdoor sports	stadiums
B	business, for office or service transactions	offices, banks, educational institutions above the 12th grade, post offices
E	educational use by ≥ 6 people through 12th grade	grade schools, high schools, day cares with more than 5 children, ages greater than 2.5 yr
F-1	factory, moderate hazard	see Code
F-2	factory, low hazard	see Code
H	hazardous—see Code	see Code
I-1	>16 ambulatory people on 24 hr basis	assisted living, group home, convalescent facilities
I-2	medical care on 24 hr basis to >5 people not capable of self-preservation	hospitals, skilled care nursing
I-3	>5 people restrained	jails, prisons, reformatories
I-4	day care for >5 adults or infants (<2.5 yr)	day care for infants
M	mercantile	department stores, markets, retail stores, drug stores, sales rooms
R-1	residential, for transient lodging	hotels and motels
R-2	residential with 3 or more units	apartments, dormitories, condominiums, convents
R-3	1 or 2 dwelling units with attached uses or child care <6, less than 24 hr care	bed and breakfast, small child care
R-4	residential assisted living where number of occupants >5 but <16	small assisted living
S	storage—see Code	see Code
U	utility—see Code	see Code
dwellings	must use *International Residential Code*	

*This table briefly summarizes the groups and examples of occupancy groups. Refer to the IBC for a complete list or check with local building officials when a use is not clearly stated or described in the code.

previously been disturbed, is part of a sloped, layered system where the layers dip into the excavation on a slope of four horizontal to one vertical or greater, or has seeping water.

Type B soils are cohesive soils with an unconfined compressive strength greater than 0.5 tons per square foot (48 kPa) but less than 1.5 (144 kPa). Examples of type B soils are angular gravel; silt; silt loam; previously disturbed soils unless otherwise classified as type C; soils that meet the unconfined compressive strength or cementation requirements of type A soils but are fissured or subject to vibration; dry unstable rock; and layered systems sloping into the trench at a slope less than four horizontal to one vertical (but only if the material would be classified as a type B soil).

Type C soils are cohesive soils with an unconfined compressive strength of 0.5 tons per square foot (48 kPa) or less. Type C soils include granular soils such as gravel, sand and loamy sand, submerged soil, soil from which water is freely seeping, and submerged rock that is not stable. Also included in this classification is material in a sloped, layered system where the layers dip into the excavation or have a slope of four horizontal to one vertical or greater.

Layered geological strata include soils that are configured in layers. Where a layered geologic structure exists, the soil must be classified on the basis of the soil classification of the weakest soil layer. Each layer may be classified individually if a more stable layer lies below a less stable layer (e.g., where a type C soil rests on top of stable rock).

Figure 83.1 Soil Type by Geotechnical Qualities

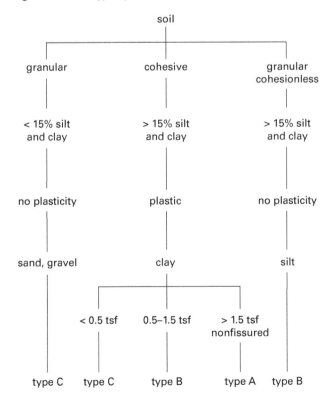

6. TRENCHING AND EXCAVATION

(OSHA 1926 Subpart P App. C)

Figure 83.2 Slope and Shield Configurations

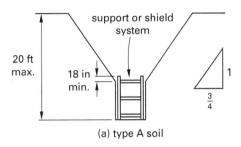

(a) type A soil

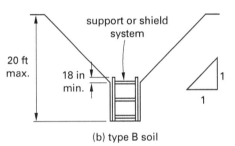

(b) type B soil

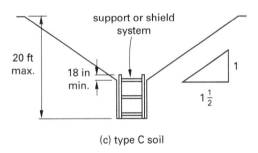

(c) type C soil

Maximum Allowable Slopes
(OSHA 1926 Subpart P App. B)

Table 83.3 Maximum Allowable Slopes

soil or rock type	maximum allowable slopes $(H{:}V)^a$ for excavations less than 20 ft deep[b]
stable rock	vertical (90°)
type A[c]	$3/4{:}1$ (53°)
type B	1:1 (45°)
type C[d]	$1^{1}/_{2}{:}1$ (34°)

[a]Numbers shown in parentheses next to maximum allowable slopes are angles expressed in degrees from the horizontal. Angles have been rounded off.
[b]Sloping or benching for excavations greater than 20 ft (6 m) deep must be designed by a registered professional engineer.
[c]A short-term maximum allowable slope of $1/2$H:1V (63°) is allowed in excavations in type A soil that are 12 ft (3.67 m) or less in depth. Short-term maximum allowable slopes for excavations greater than 12 ft (3.67 m) in depth must be $3/4$H:1V (53°).
[d]These slopes must be reduced 50% if the soil shows signs of distress.

Source: OSHA 1926 Subpart P App. B

Sloping and Benching
(OSHA 1926 Subpart P App. B)

(See Fig. 83.3 and Fig. 83.4.)

10. ELECTRICAL SAFETY

(See Table 83.4.)

11. POWER LINE HAZARDS

$$\text{line clearance} = 3 \text{ m} + (10.2 \text{ mm})(V_{kV} - 50 \text{ kV})$$
$$[\text{SI}] \quad 83.2(a)$$

$$\text{line clearance} = 10 \text{ ft} + (0.4 \text{ in})(V_{kV} - 50 \text{ kV})$$
$$[\text{U.S.}] \quad 83.2(b)$$

Figure 83.3 Excavations by Soil Type

simple slope–general

simple slope—short term
(less than 24 hours)

simple bench

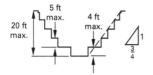

multiple bench

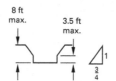

unsupported vertical sided lower portion
(maximum 8 ft in depth)

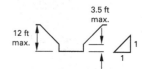

unsupported vertical sided lower portion
(maximum 12 ft in depth)

type A soil

simple slope

single bench
(allowed in cohesive soil only)

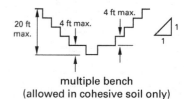

multiple bench
(allowed in cohesive soil only)

type B soil

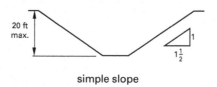

simple slope

type C soil

Figure 83.4 Excavations in Layered Soils

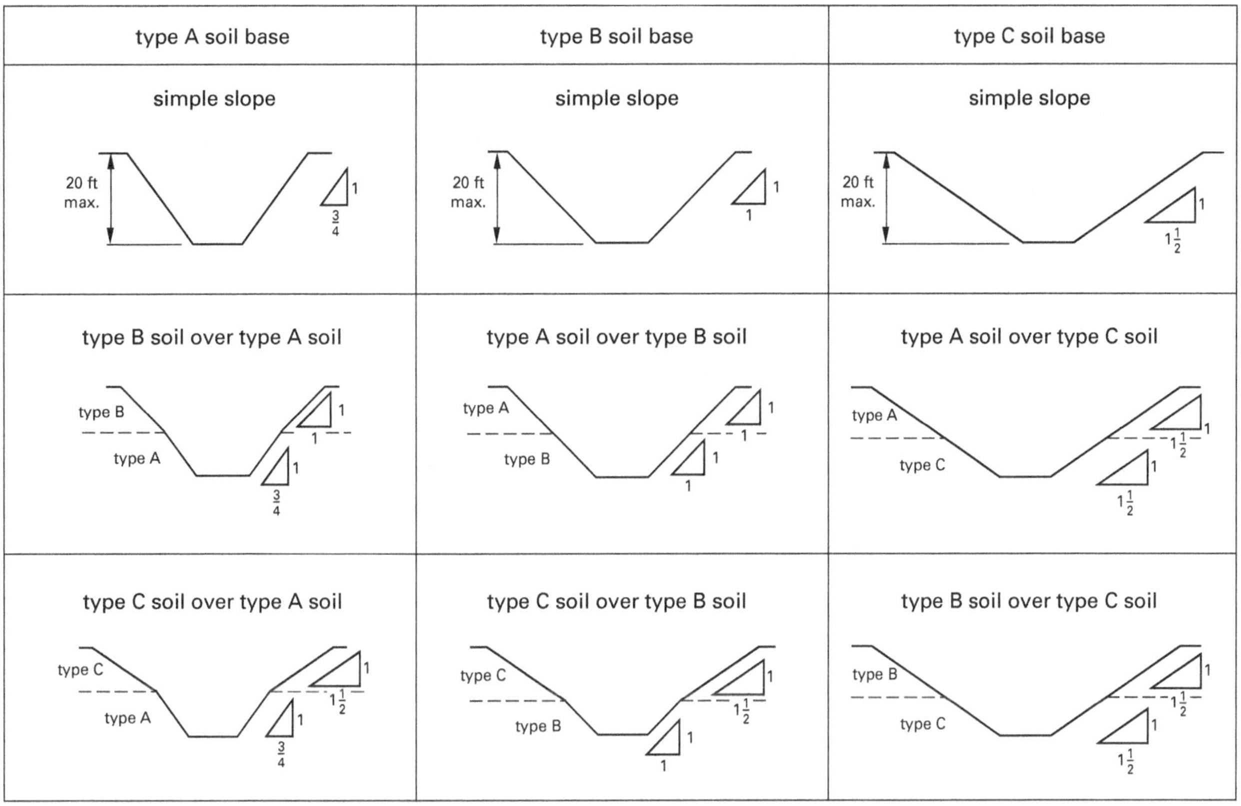

Table 83.4 Effects of Current on Humans

current level	probable effect on human body
1 mA	Perception level. Slight tingling sensation. Still dangerous under certain conditions.
5 mA	Slight shock felt; not painful, but disturbing. Average individual can let go. However, strong involuntary reactions to shocks in this range may lead to injuries.
6–16 mA	Painful shock, begin to lose muscular control. Commonly referred to as the freezing current or "let go" range.
17–99 mA	Extreme pain, respiratory arrest, severe muscular contractions. Individual cannot let go. Death is possible.
100–2000 mA	Ventricular fibrillation (uneven, uncoordinated pumping of the heart). Muscular contraction and nerve damage begins to occur. Death is likely.
> 2000 mA	Cardiac arrest, internal organ damage, and severe burns. Death is probable.

13. NIOSH LIFTING EQUATION

$$\text{LI} = \frac{L}{\text{RWL}} \qquad 83.3$$

$$\text{RWL} = (\text{LC})(\text{HM})(\text{VM})(\text{DM})(\text{AM})(\text{FM})(\text{CM}) \qquad 83.4$$

14. NOISE

(See Table 83.5.)

$$D = \sum \frac{C_i}{T_i} \times 100\% \qquad 83.5$$

$$T_i = \frac{8 \text{ hr}}{2^{(L_i-90)/5}} \qquad 83.6$$

$$\text{TWA} = 90 + 16.61 \log_{10} \frac{D}{100\%} \qquad 83.7$$

$$\Delta L_{\text{dBA}} = \frac{\text{NRR} - 7}{2} \quad \text{[formable earplugs]} \qquad 83.8$$

$$\Delta L_{\text{dBC}} = \text{NRR} - 7 \quad \text{[formable earplugs]} \qquad 83.9$$

Table 83.5 *Typical Permissible Noise Exposure Levels**

sound level (dBA)	exposure (hr/day)
90	8
92	6
95	4
97	3
100	2
102	$1\frac{1}{2}$
105	1
110	$\frac{1}{2}$
115	$\frac{1}{4}$ or less

*without hearing protection

Source: OSHA 1910.95, Table G-16

Systems, Management, and Professional

CERM Chapter 86
Project Management, Budgeting, and Scheduling

5. ACTIVITY-ON-NODE NETWORKS

$$\text{float} = LS - ES = LF - EF \qquad 86.1$$

6. SOLVING A CPM PROBLEM

ES	EF
LS	LF

key

ES: Earliest Start

EF: Earliest Finish

LS: Latest Start

LF: Latest Finish

step 1: Place the project start time or date in the **ES** and **EF** positions of the start activity. The start time is zero for relative calculations.

step 2: Consider any unmarked activity, all of whose predecessors have been marked in the **EF** and **ES** positions. (Go to step 4 if there are none.) Mark in its **ES** position the largest number marked in the **EF** position of those predecessors.

step 3: Add the activity time to the **ES** time and write this in the **EF** box. Go to step 2.

step 4: Place the value of the latest finish date in the **LS** and **LF** boxes of the finish mode.

step 5: Consider unmarked predecessors whose successors have all been marked. Their **LF** is the smallest **LS** of the successors. Go to step 7 if there are no unmarked predecessors.

step 6: The **LS** for the new node is **LF** minus its activity time. Go to step 5.

step 7: The float for each node is **LS** − **ES** or **LF** − **EF**.

step 8: The critical path encompasses nodes for which the float equals **LS** − **ES** from the start node. There may be more than one critical path.

8. STOCHASTIC CRITICAL PATH MODELS

$$\mu = \tfrac{1}{6}(t_{\min} + 4t_{\text{most likely}} + t_{\max}) \qquad 86.2$$

$$\sigma = \tfrac{1}{6}(t_{\max} - t_{\min}) \qquad 86.3$$

$$p\{\text{duration} > D\} = p\{t > z\} \qquad 86.4$$

$$z = \frac{D - \mu_{\text{critical path}}}{\upsilon_{\text{critical path}}} \qquad 86.5$$

12. EARNED VALUE METHOD

$$CV = BCWP - ACWP \qquad 86.6$$

$$SV = BCWP - BCWS \qquad 86.7$$

$$CPI = \frac{BCWP}{ACWP} \qquad 86.8$$

$$SPI = \frac{BCWP}{BCWS} \qquad 86.9$$

$$ETC = BAC - BCWP \qquad 86.10$$

$$EAC = ACWP + ETC \qquad 86.11$$

CERM Chapter 87
Engineering Economic Analysis

11. SINGLE-PAYMENT EQUIVALENCE

$$F = P(1 + i)^{n} \qquad 87.2$$

$$P = F(1 + i)^{-n} = \frac{F}{(1 + i)^{n}} \qquad 87.3$$

12. STANDARD CASH FLOW FACTORS AND SYMBOLS

Table 87.1 Discrete Factors for Discrete Compounding

factor name	converts	symbol	formula
single payment compound amount	P to F	$(F/P, i\%, n)$	$(1+i)^n$
single payment present worth	F to P	$(P/F, i\%, n)$	$(1+i)^{-n}$
uniform series sinking fund	F to A	$(A/F, i\%, n)$	$\dfrac{i}{(1+i)^n - 1}$
capital recovery	P to A	$(A/P, i\%, n)$	$\dfrac{i(1+i)^n}{(1+i)^n - 1}$
uniform series compound amount	A to F	$(F/A, i\%, n)$	$\dfrac{(1+i)^n - 1}{i}$
uniform series present worth	A to P	$(P/A, i\%, n)$	$\dfrac{(1+i)^n - 1}{i(1+i)^n}$
uniform gradient present worth	G to P	$(P/G, i\%, n)$	$\dfrac{(1+i)^n - 1}{i^2(1+i)^n} - \dfrac{n}{i(1+i)^n}$
uniform gradient future worth	G to F	$(F/G, i\%, n)$	$\dfrac{(1+i)^n - 1}{i^2} - \dfrac{n}{i}$
uniform gradient uniform series	G to A	$(A/G, i\%, n)$	$\dfrac{1}{i} - \dfrac{n}{(1+i)^n - 1}$

24. CHOICE OF ALTERNATIVES: COMPARING ONE ALTERNATIVE WITH ANOTHER ALTERNATIVE

Capitalized Cost Method

$$\text{capitalized cost} = \text{initial cost} + \frac{\text{annual costs}}{i} \quad 87.19$$

$$\text{capitalized cost} = \text{initial cost} + \frac{\text{EAA}}{i}$$

$$= \text{initial cost} + \frac{\text{present worth}}{\text{of all expenses}} \quad 87.20$$

25. CHOICE OF ALTERNATIVES: COMPARING AN ALTERNATIVE WITH A STANDARD

Benefit-Cost Ratio Method

$$B/C = \frac{\Delta \, \substack{\text{user} \\ \text{benefits}}}{\Delta \, \substack{\text{investment} \\ \text{cost}} + \Delta \, \text{maintenance} - \Delta \, \substack{\text{residual} \\ \text{value}}} \quad 87.21$$

26. RANKING MUTUALLY EXCLUSIVE MULTIPLE PROJECTS

$$\frac{B_2 - B_1}{C_2 - C_1} \geq 1 \quad \text{[alternative 2 superior]} \quad 87.22$$

30. TREATMENT OF SALVAGE VALUE IN REPLACEMENT STUDIES

$$\begin{aligned} \text{EUAC (defender)} = & \text{ next year's maintenance costs} \\ & + i(\text{current salvage value}) \\ & + \text{current salvage} \\ & - \text{next year's salvage} \end{aligned} \quad 87.23$$

35. DEPRECIATION BASIS OF AN ASSET

$$\text{depreciation basis} = C - S_n \quad 87.24$$

36. DEPRECIATION METHODS

Straight-Line Method

$$D = \frac{C - S_n}{n} \qquad 87.25$$

Sum-of-the-Years' Digits Method

$$T = \tfrac{1}{2}n(n+1) \qquad 87.27$$

$$D_j = \frac{(C - S_n)(n - j + 1)}{T} \qquad 87.28$$

Double Declining Balance Method

$$D_{\text{first year}} = \frac{2C}{n} \qquad 87.29$$

$$D_j = \frac{2\left(C - \sum\limits_{m=1}^{j-1} D_m\right)}{n} \qquad 87.30$$

$$d = \frac{2}{n} \quad \begin{bmatrix} \text{double declining} \\ \text{balance} \end{bmatrix} \qquad 87.31$$

$$D_j = dC(1 - d)^{j-1} \qquad 87.32$$

Statutory Depreciation Systems

$$D_j = C \times \text{factor} \qquad 87.33$$

Table 87.4 *Representative MACRS Depreciation Factors**

| | depreciation rate for recovery period, n | | | |
year, j	3 years	5 years	7 years	10 years
1	33.33%	20.00%	14.29%	10.00%
2	44.45%	32.00%	24.49%	18.00%
3	14.81%	19.20%	17.49%	14.40%
4	7.41%	11.52%	12.49%	11.52%
5		11.52%	8.93%	9.22%
6		5.76%	8.92%	7.37%
7			8.93%	6.55%
8			4.46%	6.55%
9				6.56%
10				6.55%
11				3.28%

*Values are for the "half-year" convention. This table gives typical values only. Since these factors are subject to continuing revision, they should not be used without consulting an accounting professional.

Production or Service Output Method

$$D_j = (C - S_n)\left(\frac{\text{actual output in year } j}{\text{estimated lifetime output}}\right) \qquad 87.34$$

Sinking Fund Method

$$D_j = (C - S_n)(A/F, i\%, n)(F/P, i\%, j - 1) \qquad 87.35$$

39. BOOK VALUE

For the straight-line depreciation method, the book value at the end of the jth year, after the jth depreciation deduction has been made, is

$$\text{BV}_j = C - \frac{j(C - S_n)}{n} = C - jD \qquad 87.36$$

For the sum-of-the-years' digits method, the book value is

$$\text{BV}_j = (C - S_n)\left(1 - \frac{j(2n + 1 - j)}{n(n+1)}\right) + S_n \qquad 87.37$$

For the declining balance method, including double declining balance, the book value is

$$\text{BV}_j = C(1 - d)^j \qquad 87.38$$

For the sinking fund method, the book value is calculated directly as

$$\text{BV}_j = C - (C - S_n)(A/F, i\%, n)(F/P, i\%, j) \qquad 87.39$$

For any method by successive subtractions, the book value is

$$\text{BV}_j = C - \sum_{m=1}^{j} D_m \qquad 87.40$$

42. BASIC INCOME TAX CONSIDERATIONS

$$t = s + f - sf \qquad 87.41$$

46. RATE AND PERIOD CHANGES

$$\phi = \frac{r}{k} \qquad \qquad 87.50$$

$$i = (1 + \phi)^k - 1$$
$$= \left(1 + \frac{r}{k}\right)^k - 1 \qquad 87.51$$

48. PROBABILISTIC PROBLEMS

$$\mathcal{E}\{\text{cost}\} = p_1(\text{cost } 1) + p_2(\text{cost } 2) + \cdots \qquad 87.52$$

49. WEIGHTED COSTS

$$C_{\text{weighted}} = w_1 C_1 + w_2 C_2 + \cdots + w_N C_N \qquad 87.53$$

$$w_j = \frac{A_j}{\sum_{i=1}^{N} A_i} \quad [\text{weighting by area}] \qquad 87.54$$

52. ACCOUNTING PRINCIPLES

$$\text{assets} = \text{liability} + \text{owner's equity} \qquad 87.55$$

55. BREAK-EVEN ANALYSIS

$$C = f + aQ \qquad 87.68$$

$$R = pQ \qquad 87.69$$

$$Q^* = \frac{f}{p - a} \qquad 87.70$$

58. INFLATION

$$i' = i + e + ie \qquad 87.72$$

Figure 87.13 *Break-Even Quality*

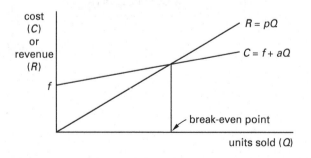

60. FORECASTING

Forecasts by Moving Averages

$$F_{t+1} = \frac{1}{n} \sum_{m=t+1-n}^{t} D_m \qquad 87.86$$

Forecasts by Exponentially Weighted Averages

$$F_{t+1} = \alpha D_t + (1 - \alpha) F_t \qquad 87.87$$

61. LEARNING CURVES

$$T_n = T_1 n^{-b}$$
$$\int_{n_1}^{n_2} T_n \, dn \approx \left(\frac{T_1}{1-b}\right)\left(\left(n_2 + \tfrac{1}{2}\right)^{1-b} - \left(n_1 - \tfrac{1}{2}\right)^{1-b}\right)$$
$$87.89$$

$$T_{\text{ave}} = \frac{\int_{n_1}^{n_2} T_n \, dn}{n_2 - n_1 + 1} \qquad 87.90$$

$$b = \frac{-\log_{10} R}{\log_{10}(2)} = \frac{-\log_{10} R}{0.301} \qquad 87.91$$

62. ECONOMIC ORDER QUANTITY

$$t^* = \frac{Q}{a} \qquad 87.92$$

$$H = \tfrac{1}{2} Q h t^* = \frac{Q^2 h}{2a} \qquad 87.93$$

$$C_t = \frac{aK}{Q} + \frac{hQ}{2} \qquad 87.94$$

$$Q^* = \sqrt{\frac{2aK}{h}} \qquad 87.95$$

$$t^* = \frac{Q^*}{a} \qquad 87.96$$

Figure 87.18 *Inventory with Instantaneous Reorder*

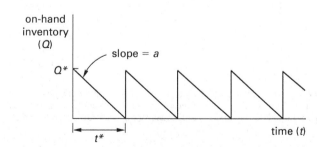

Index

A

AASHTO
  method, rigid pavement design, 109
  required curve lengths, 115 (tbl)
  soil classification, 49
Abbreviation, stake marking, 119 (tbl)
Absolute
  viscosity, 19
  volume method, 75
Acceleration
  uniform, 102
  uniform, formulas, 102 (tbl)
Accident
  analysis, 107
  data, 107
  modeling vehicle, 107
Accounting principles, 128
ACI
  moment coefficient, 74 (fig)
  shear coefficient, 74
Acidity, 37
Activated sludge, 43
Active earth pressure, 54
Activity-on-node networks, 125
Actual air, 36
  combustion, 37
Aerated lagoon, 42
Aeration, 38
  cost, 45
  power, 45
  tank, 44
Affinity law, 27
Aggregate factor, lightweight
    concrete, 75 (tbl)
Air
  actual, 36
  actual combustion, 37
  atmospheric, 36
  composition, dry, 36 (tbl)
  excess, 36
  excess combustion, 37
  stoichiometric, 36
  stripping, 47
Algebra, 9
  linear, 9
Alkalinity, 37, 38 (fig)
Allowable
  compressive strength, steel column, 92
  pipe load, 61
  stress design, 69
Alternative choice, 126
Analysis
  accident, 107
  anchored bulkhead, 58
  braced excavation, 58
  break-even, 128
  capacity, 104
  error, 109
  flexible bulkhead, 58
  of prestressed beams, 86
  of traffic accidents, 107
  second-order, 82
Analytic geometry, 12
Anchorage shear reinforcement, 81
Anchored bulkhead, 58
  analysis, 58
  design, 58

B

Backfilled trench, pipe in, 61 (fig)
Backwashing, filter, 40
Balance
  line, 118 (fig)
  point, 118 (fig)
Balanced traverse, 111
Banking, roadway, 103
Bar area, 79 (tbl)
Base plate, column, 93
Basic elements of design, 69
Basin, plain sedimentation, 43
Basis,
  calculation, dry and wet, 76 (tbl)
  depreciation, asset, 126
Beam (see also type)
  balanced condition, 77 (fig)
  bearing plate, 88
  bearing plate, nomenclature, 90 (fig)
  bending stress, 68
  bending stress distribution, 69 (fig)
  -column, 94
  -column design, 94

Angle
  approximation, small, 11
  bearing, 111 (fig)
  miscellaneous formula, 11
Application reliability, 16
Applied load, pressure, 59
Approximate method, 74
Approximation, small angle, 11
Aquifer, 32
  characteristics, 32
Archaic function, 11
Area, 5
  bar, 79 (tbl)
  centroid, 64
  effective net, 91
  first moment, 66
  gross, 90
  irregular, 12 (fig), 111
  irregular boundaries, 12, 111
  mensuration, 5
  method, moment, 69
  moment of inertia, 66
  net, 90
  properties of, 64
  traverse, 111
  under standard normal curve, 14 (tbl)
Asphalt concrete, 108
At-rest soil pressure, 55
Atmospheric air, 36
Atomic
  number, table, 34
  weight, table, 34
Atterberg limit tests, 49
Auger hole, 49
Available
  flexural strength, 96
  moment versus unbraced length, 87 (fig)
Average end area method, 118
Axial
  compression, 99
  tensile strength, 90
  tensile stress, 90
Axis theorem, parallel, 66

concrete, design, 78
concrete, shear stress, 80
continuous, 88
deflection, 69, 74
depth, concrete, minimum, 79
shear stress, 68
steel, analysis flowchart, 89 (fig)
T-, design of, 80
thickness, minimum, 80 (tbl)
unsymmetrical bending, 88
width, 77
width, minimum, 77 (tbl)
Bearing
  angle, 111 (fig)
  capacity, 51
  capacity factor, 52 (tbl)
  capacity, clay, 52
  capacity, sand, 52
  capacity, water table, 53
  plate, beam, 88
  stiffener, 96 (fig)
  strength, available, 96
  stress, allowable, 96
Benching, excavations, 122 (fig)
Bend, pipe, 25 (fig)
Bending
  biaxial, 100
  strength, steel beam, 87
  stress, beam, 68, 69 (fig)
  unsymmetrical, beam, 88
Benefit-cost ratio, 126
Bernoulli equation, 21
Biaxial
  bending, 100
  loading, 68
Billion, parts per, 33
Binomial distribution, 13
Biochemical oxygen demand, 42
Biofiltration, 47
Block shear strength, 91
BOD, 42
  escaping treatment, soluble, 43
  seeded, 42
  soluble, 43
Body
  rigid, 103
  solid, 101
Bolt
  connection, 71
  preload, 72
Bolted connection reduction values, 91 (tbl)
Book value, 127
Boussinesq's equation, 59
Braced
  column, 82, 83
  cut, clay, 57
  cut, sand, 57
  cut, stability, 58
  excavation, 58
  excavation, analysis, 58
  excavation, design, 58
  excavation, stability, clay, 58
  excavation, stability, sand, 58
Bracing, lateral, 87
Braking
  distance, 107
  rate, 107

Break-even
 analysis, 128
 quality, 128 (fig)
Broad
 -crested weir, 29
 fill, 61
Buckling
 load, 83
 local, 93
Budgeting, 125
Building code, 120
Built-up section, 94
Bulk modulus, 19
Bulkhead
 anchored, 58
 flexible, 58
Buoyancy, 21
 force, 21
Buried pipe, loads on, 60

C
Cable, 64
 catenary, 64, 64 (fig)
 parabolic, 64 (fig)
California bearing test ratio, 51
Cantilever retaining wall
 analysis, 56
 design, 56
Capacity, 32
 analysis, 104
 bearing, 51
 bearing, clay, 52
 bearing, sand, 52
 highway, 104
 landfill, 46
 pile, 56
 pile group, 57
 point-bearing, 57
 road, 104
 skin-friction, 57
 specific, 32
Capitalized cost, 126
Cartesian unit vector, 10 (fig)
Cash flow factor, 126
Catenary cable, 64, 64 (fig)
Cavitation coefficient, 27
CBR, 51
 test, 51
Center of
 gravity, 101
 rotation, 97
Central tendency, 16
 measures of, 16
Centrifugal pump, 26
Centripetal force, 102
Centroid
 area, 64
 line, 66
Chamber, grit, 42
Change
 period, 128
 rate, 128
Channel
 flow, open, 28
 most efficient, 29
 nonrectangular, 30
 rectangular, 29, 30
 trapezoidal, 29
Characteristics
 aquifer, 32
 electrostatic precipitator, 47
 impeller, 39
Characterizing rock mass quality, 51
Chart, influence, 59
Chemical
 elements, table, 35
 symbol, 35
Chemistry, inorganic, 33
Chezy
 equation, 28
 -Manning equation, 28

Choice, alternative, 126
Chord offset, 113 (fig)
 curve layout, 113
Churchill formula, 42
Circle, 13
 formula, 5
 segment, 5
Circular
 shaft, 72
 shaft, design, 72
 transcendental function, 10
Clarification, plain sedimentation, 43
Classification
 culvert flow, 31 (fig)
 soil, 49, 120
Clay
 bearing capacity, 52
 braced cut, 57
 condition, 59
 consolidation, 59
 consolidation curve, 59 (fig)
 cut in soft, 58 (fig)
 cut in stiff, 58 (fig)
 normally consolidated, 59
 overconsolidated, 59 (fig)
 raft on, 53
 slope stability, 60
 stability of braced excavation, 58
 varved, 59
Clearance, power line, 122
Climatology, 31
Coagulant dose, 39
Code, building, 120
Coefficient
 ACI, shear, 74
 cavitation, 27
 drag, 25
 end restraint, 70 (tbl)
 of friction, 106
 of restitution, 103, 104
 of variation, 16
 valve flow, 23
Column
 analysis, steel, 92
 base plate, 93, 94 (fig)
 braced, 82, 83
 eccentrically loaded, 70
 footing, 84
 intermediate, 70
 slender, 70
 spiral, 82
 tied, 82
 unbraced, 82, 83
Combined
 shear and tension connection, 97 (fig)
 stress, 68 (fig)
 stresses, 68
Combustible gas, 36
Combustion, 34
 efficiency, 37
 heat of, 36
 loss, 37
 reaction, ideal, 36 (tbl)
 reactions, 36
 temperature, 37
Comfort, curve length, 115
Compact section, 87
Component, moment, 63
Composite
 beam, deflection design model, 96 (fig)
 beam, steel, 96
 concrete, 86
 structure, 69
Composition, dry air, 36 (tbl)
Compound curve, 111
Compounding, discrete, 126 (tbl)
Compression
 axial, 94, 99
 flexural, 94

Compressive strength
 concrete, 74
 steel column, 92
 unconfined, 51
Computed quantity, error, 109
Concentrated force, 63
Concentration
 stress, 68
 time of, 31
 units of, 33
Concentric tension connection, 97
Concrete
 asphalt, 108
 beam, 76
 beam, shear stress, 80
 components, properties, 75 (tbl)
 compressive strength, 74
 footing, 84
 lightweight aggregate factor, 75 (tbl)
 mixing, 75
 modulus of rupture, 75
 properties of, 74
 shear strength, 80
 slab, effective width, 96
Condition
 clay, 59
 maximum moment, 78 (fig)
Conditions, no-stirrup, 81
Conduit, buried, 60
Cone penetrometer test, 49
Confidence
 level, values of z for various, 16 (tbl)
 limit, 16
 limit, one-tail, 16 (tbl)
 limit, two-tail, 16 (tbl)
Confined stream, 25
 in pipe bend, 25
Conical pile, 118 (fig)
Connection
 bolt, 71
 eccentric, 72
 eccentric, welded, 97
 reduction value, 91 (tbl)
 rivet, 71
 tension, 96
Conservation of mass, 22
Consolidation
 clay, 59
 curve, clay, 59 (fig)
 parameters, 59, 60 (fig)
 primary, 60
 rate, 60
 rate, primary, 60
 secondary, 60
 test, 51
Constant
 equilibrium, 34
 force, work, 16 (fig)
 ionization, 34
Construction, 117
 earthwork, 117
Continuous
 beam, portion, 73 (fig)
 plastic design, 88
Contraction
 loss, 23
 sudden, 23
Control
 odor, 40
 taste, 40
 zones, temporary traffic, 106
Convention, sign, 68 (fig)
Conversion
 formulas, power, 17 (tbl)
 water quality, 37
Coordinate, fluid stream, 23 (fig)
Coordinates, method of, 111
Cosine, 10

Cost
   aeration, 45
   -benefit ratio, 126
   capitalized, 126
   of electricity, 26
   weighted, 128
Coulomb theory, 54
Council, National Research, 43
Couple, 63 (fig)
Cover, steel, 77
CPM problem, 125
Cracked moment of inertia, 78, 79 (fig)
Cracking, 78
   calculation parameter, 79 (fig)
Creep, 85
Crest curve, 115
   design, 115
Critical
   depth, 30
   flow, 30
   path, 125
   section, one-way shear, 84 (fig)
   section, two-way shear, 85 (fig)
Cross
   product, vector, 10 (fig)
   section, most efficient, 29
   section, trapezoidal, 29 (fig)
Culvert, 23
   flow, 30
   flow classification, 31 (fig)
   flow, type of, 30
   simple pipe, 23 (fig)
Current, electrical, effect on
      humans, 123 (tbl)
Curve
   area under standard normal, 14 (tbl)
   compound, 111
   crest, 115
   deflection angle, 112 (fig)
   degree of, 112
   elements, horizontal, 112 (fig)
   Euler's, 70 (fig)
   horizontal, 111
   horizontal, through point, 113 (fig)
   layout, 112
   layout, chord offset, 113
   layout, tangent offset, 113
   learning, 128
   length, AASHTO required, 115 (tbl)
   length, comfort, 115
   length, spiral, 116
   length, vertical, 115
   oxygen sag, 42 (fig)
   sag, 115
   spiral, 111, 115, 116 (fig)
   stationing, 112
   system, 27
   vertical, 114
   vertical, obstruction, 115 (fig)
   vertical, symmetrical parabolic, 114 (fig)
   vertical, through point, 115
Cut
   braced, clay, 57
   braced, sand, 57
   in sand, 57 (fig)
   in soft clay, 58 (fig)
   in stiff clay, 58 (fig)
Cycle length, 105
Cyclone separator, 47
Cylinder, thick-walled, 71
Cylindrical tank, thin-walled, 70

D

Dam, 21 (fig)
   forces on, 20
Darcy equation, 22
Darcy's law, 32
Data accident, 107
Deceleration rate, 107
Deep foundation, 56

Deflection, 86
   angle, curve, 112 (fig)
   beam, 69, 74
   concrete beam, long-term, 79
   design, model for composite
      beam, 96 (fig)
   two-way slabs, 82
Deformation
   elastic, 67, 73
   linear, 74
   thermal, 67
Degree, 10
   of curve, 112
   of indeterminacy, 73
Demand
   fire fighting, 40
   water, 40
Density, 19
   optical, 47
   relationship, 104
Deoxygenation, 42
Depreciation, 126, 127
   basis, asset, 126
   double declining balance, 127
   method, 126, 127
   statutory, 127
   straight-line, 127
   sum-of-the-years' digits, 127
Depth
   concrete beam, minimum, 79
   critical, 30
   hydraulic, 28
   normal, 28
   pressure at, 20
   -thickness ratio, 94
Design
   anchored bulkhead, 58
   beam-column, 94
   bearing stiffener, 95
   braced excavation, 58
   crest curve, 115
   criteria, 76
   flexible bulkhead, 58
   flexible pavement, 107
   footing, 53
   gravel screen, 32
   Hveem mix, 108
   of steel beams, 88
   porous filter, 32
   strength, 76
   strength, flexural, 87
   traffic, 109
Determinant, 9
Determinate
   statics, 63
   truss, 64
Detour, 106
Development length, 85
   masonry, 98
   of flexural reinforcement, 85
Deviation, standard, 16
Dew point, flue gas, 36
Dewatering, sludge, 46
Diagram
   influence, 74
   mass, 118
Diameter
   equivalent, 21 (tbl)
   hydraulic, 21
Differential equation, 13
Dilution purification, 42
   response to, 42
Direct
   central impact, 103 (fig)
   design method, 81
   leveling, 110 (fig)
   shear test, 51
Direction, 110
   specification, 110
Discharge
   from tank, 23 (fig)
   velocity, 23, 32

Discount factor, discrete
      compounding, 126 (tbl)
Discrete compounding, discount
      factor, 126 (tbl)
Disk, drag on, 25
Dispersion, 16
   measures of, 16
Dissolved oxygen, wastewater, 41
Distance, 101
   between geometric figures, 12
   between points, 12
   braking, 107
   clear, 80
   double meridian, 111
   haul, 118 (fig)
   measurement, 110
   passing sight, 115
   sight, 115
   skidding, 107
   stopping, 106, 114
   stopping sight, 114
Distributed load, 63
Distribution
   binomial, 13
   normal, 13
   Poisson, 13
   pressure, 20
   water supply, 38
Dobbins formula, 41
Dose
   coagulant, 39
   noise, permissible, 124 (tbl)
Dot product, vector, 10 (fig)
Double
   angle formula, 11
   declining balance depreciation, 127
   integration method, 69
   meridian distance, 111
Doubly reinforced
   beam, 81
   section, 81
Drag, 25
   coefficient, 25
   on disks, 25
   on spheres, 25
Drawdown
   in aquifers, well, 32
   well, 32
Drinking water demand, 40
Driver, pile, 56
Dry air, composition, 36 (tbl)
Ductility, 67
Dummy unit load method, 74
Dupuit equation, 32
Dynamics
   fluid, 22
   vehicle, 106

E

e-log p curve, 51 (fig)
Earliest
   finish, 125
   start, 125
Earned value method, 125
Earth pressure, 53
   active, 54
   active and passive, 54 (fig)
   passive, 54
Earthwork construction, 117
Eccentric
   load, column, 70
   load, footing, 53
   load, rectangular footing, 53
   loading, 69
   loading, axial member, 69 (fig)
Eccentricity
   column, 82
   design, small, 82
   normal to plane of faying surface, 97 (fig)

INDEX - E

Economic
    analysis, engineering, 125
    appraisal, HSM, 107
    order quantity, 128
    ranking, 126
Economy, fuel, 106
Effective
    length, 92
    length factor, 92 (tbl)
    net area, 91
    slab width, composite member, 86
    stress, 55
    width, 96
    width, concrete slab, 96
Efficiency, 17
    combustion, 37
    process, 44
    pump, 26
Efficient
    channel, most, 29
    cross section, most, 29
Elastic
    deformation, 67, 73
    method, 97
    method, eccentric connection, 97
Electric motor, 27
Electrical
    current, effect on humans, 123 (tbl)
    safety, 123 (tbl)
Electricity, cost of, 26
Electrostatic precipitator, 47
    characteristics, 47
Element
    horizontal curve, 112 (fig)
    table, 34
Elevation measurement, 110
Ellipse formula, 5
End
    area method, average, 118
    restraint coefficient, 70 (tbl)
Energy, 16, 74
    grade line, 22
    gradient, 28
    internal, 17
    kinetic, 21
    loss, 22
    loss, turbulent, 22
    of a mass, internal, 17
    of a mass, kinetic, 17
    of a mass, potential, 17
    of a mass, pressure, 21
    potential, 17, 21
    specific, 22, 29
    spring, 17
    strain, 67, 69
Engineering
    economic analysis, 125
    economy, 125
Enlargement
    loss, 23
    sudden, 23
Enthalpy of reaction, 34
Environmental
    pollutants, 47
    remediation, 47
Equation
    Bernoulli, 21
    Boussinesq's, 59
    Chezy, 28
    Chezy-Manning, 28
    Darcy, 22
    differential, 13
    Dupuit, 32
    general bearing capacity, 51
    Harmon's, 41
    Hazen-Williams, 22
    National Research Council, 43
    Niosh lifting, 123
    quadratic, 9
    rational, 31
    Theis, 33
    Thiem, 32

three-moment, 73
    Torricelli's, 23
    TSS, 41
    uniform flow, 28
    Velz, 43
    Webster's, 105
Equilibrium
    conditions of, 63
    constant, 34
Equipment
    placement and paving, 107, 108
    wastewater treatment, 42
Equivalence, single-payment, 125
Equivalent
    diameter, 21 (tbl)
    fluid height, 20
    weight, 33
    weight, milligram, 33
Error
    analysis, 109
    in computed quantity, 109
ESAL, 109
Escaping treatment, 43
Euler buckling, 91
Euler's
    curve, 70 (fig)
    theory, 91
Excavation, 57, 121
    benching, 122 (fig)
    braced, 58
    by soil type, 122 (fig)
    layered soil, 122 (fig), 123 (fig)
    safety, 121
    temporary, 57
Excess air, 36
    combustion, 37
Exchange, ion, 40
Exponent, 9
    rules, 9
Exponentially weighted averages,
        forecasts, 128
Exposure level, noise, 124 (tbl)

F
Factor (*see also type*)
    bearing capacity, 52 (tbl)
    cash flow, 126
    discount, 126 (tbl)
    effective length, 92 (tbl)
    HSM crash modification, 107
    Meyerhof, 52 (tbl)
    peak hourly, 104
    shrinkage, 117
    size, 85
    swell, 117
    Terzaghi, 52 (tbl)
    truck, 109
    Vesic, 52 (tbl)
Factored load, 76
Fastener
    allowable, 96
    stress, static loading, 96 (tbl)
Fill, broad, 61
Fillet weld, 72, 97 (fig)
    size, minimum, 97 (tbl)
Filter
    backwashing, 40
    porous, 32
    trickling, 43
Filtration, 40
Finish
    earliest, 125
    latest, 125
Fire fighting demand, 40
First moment, 66
    of area, 66
Fit, interference, 71
Fixed-end moment, 73
Flame temperature, 37
Flange plate, girder, 95
Flanged beam, web, 68 (fig)

Flat friction, 102
Flexible
    bulkhead, 58
    bulkhead, analysis, 58
    bulkhead, design, 58
    pavement design, 107
    pipe, 61
Flexural design strength, 87
Flexure, 99
Float, 125
Flood, 31
Flow
    coefficient, valve, 23
    critical, 30
    culvert, 30
    culvert, classification, 31 (fig)
    culvert, type of, 30
    measurement, weir, 29
    model, plug, 44
    net, seepage, 33
    open channel, 28
    open channel, parameters, 28
    parameter, fluid, 21
    path, hydrostatic pressure, 33
    plug, 44
    relationship, 104
    unsteady, 33
Flowchart, analysis, 89 (fig)
Flue gas, dew point, 36
Fluid
    dynamics, 22
    flow parameters, 21
    height equivalent to pressure, 20
    jet, 23
    mixture problem, 13 (fig)
    pressure, 20
    properties, 19
    statics, 20
    stream coordinates, 23 (fig)
Flume, Parshall, 29
Fluoridation, 40
Footing, 53
    column, 84
    concrete, 84
    design, 53
    eccentric load, rectangular, 53
    overturning moment, 53 (fig)
    wall, 84
Force
    about a point, moment, 63
    buoyancy, 21
    centripetal, 102
    component and direction angle, 63 (fig)
    concentrated, 63
    gravitational, 26
    hydrostatic, 20
    inertial, 26
    normal, 102
    surface tension, 26
    viscous, 26
    web, 88
Forecast by
    exponentially weighted averages, 128
    moving averages, 128
Forecasting, 128
Form
    general, 12
    intercept, 12
    normal, 12
    point-slope, 12
    polar, 12
    slope-intercept, 12
    two-point, 12
Formality, 33
Formula
    Churchill, 42
    Dobbins, 11
    double-angle, 11
    half-angle, 11
    Marston's, 60
    miscellaneous, angle, 11
    O'Connor, 41

prismoidal, 118
  two-angle, 11
  uniform acceleration, 102 (tbl)
Formwork, lateral pressure, 75
Foundation, 51
  deep, 56
  shallow, 51
Fraction
  gravimetric, 33
  ionized, 34
  mole, 33
Frame section properties, 82
Freeway, 104
Friction
  coefficient of, 106
  factor chart, Moody, 22 (fig)
  flat, 102
  loss, turbulent, 22
Froude number, 26, 30
Frustum, 118
Fuel
  economy, 106
  moisture in, 34
Function
  archaic, 11
  circular transcendental, 10
  in a unit circle, trigonometric, 10 (fig)

G

Galvanic series, 34 (tbl)
Gas
  combustible, 36
  in liquids, solutions, 33
  landfill, 46
Gauge, pitot-static, 23, 24 (fig)
General
  bearing capacity equation, 51
  form, 12
  plane surface, pressure on, 20
  triangle, 11 (fig)
Generation, trip, 104
Geometric
  figures, distance between, 12
  mean, 16
  slope, 28
Geometry, analytic, 12
Girder
  plate, 94 (fig)
  steel bridge, 86
  web, design, 95
Grade
  line, energy, 22
  rod, 118
Gradient, energy, 28
Gravel screen, design, 32
Gravimetric fraction, 33
Gravitational force, 26
Gravity
  center of, 101
  specific, 19
Greenshield's method, 105
Grit chamber, 42
Gross area, 90
Groundwater, 32
Group
  occupancy, 120 (tbl)
  pile, 57
Growth, microbial, 41
Gyration, radius of, 101

H

Half-angle formula, 11
Hammer, water, 25 (fig)
Hardness, 37, 38 (fig)
Harmonic mean, 16
Harmon's equation, 41
Haul distance, 118 (fig)
Hazards, power line, 122

Hazen-Williams
  equation, 22
  velocity, 28
Head
  loss, 22
  loss, turbulent, 22
  net positive suction, 27
Heat
  loss, 46
  of combustion, 36
  transfer, 46
Height
  equivalent, fluid, 20
  object, 111
  of the instrument, 118
Henry's law, 33
Highway
  capacity, 104
  multilane, 104
  two-lane, 104
Hole, auger, 49
Hooke's law, 67
Horizontal
  curve, 111
  curve elements, 112 (fig)
  curve, through point, 113 (fig)
  stadia measurement, 110 (fig)
Horsepower
  hydraulic, 26 (tbl)
  water, 26
Hveem mix design, 108
Hydraulic
  depth, 28
  diameter, 21
  diameter, conduit shape, 21 (tbl)
  horsepower equation, 26 (tbl)
  jump, 30
  kilowatt equation, 26 (tbl)
  machine, 26
  power, 21, 26
  radius, 28
Hydrograph unit, 31
Hydrology, 31
Hydrostatic
  force, 20
  pressure, 20, 33
  pressure, vertical plane surface, 20 (fig)

I

Ideal combustion reaction, 36 (tbl)
Identity
  logarithm, 9
  trigonometric, 11
Impact, 103
  direct central, 103 (fig)
Impeller
  characteristics, 39
  mixer, 39
Impulse, 103
  -momentum principle, 103
  -momentum principle, open system, 103
Incineration, 34
  municipal solid waste, 47
Income tax, 127
Index, sludge volume, 43
Indirect leveling, 110 (fig)
Inertia
  area moment of, 66
  mass moment of, 101
  polar moment of, 66
Inertial
  force, 26
  separator, 47
Infiltration, 33
Inflation, 128
Inflection point, 69
Influence
  chart, 59
  diagram, 74
  line, 63
  zone of, 59

Inorganic chemistry, 33
Instantaneous reorder, inventory, 128 (fig)
Instrument, height of, 118
Intensity, rainfall, 31
Intercept, 12
  form, 12
Interference, fit, 71
Intermediate
  column, 70
  stiffener, 95 (fig)
  stiffener design, 95
Internal energy, 17
  of a mass, 17
Intersection, signalized, 105
Inventory, instantaneous reorder, 128 (fig)
Ion exchange resin, regeneration, 40
Ionization constant, 34
Ionized fraction, 34
Irregular
  area, 12 (fig), 111
  boundaries, areas, 12, 111
Irrigation water, salinity, 37

J

Jet fluid, 23
Job site safety, 120
Jump, hydraulic, 30

K

K-value, 115
Kilowatt, hydraulic, 26 (tbl)
Kinematic viscosity, 19
Kinematics, 101
Kinetic
  energy, 17, 21
  energy of a mass, 17
  reaction, 34
Kinetics, 102
  mixing, 39
  reversible reaction, 34

L

Lagoon, aerated, 42
Landfill
  capacity, 46
  gas, 46
Lateral
  bracing, 87
  support, 87
  torsional buckling, 87
Latest
  finish, 125
  start, 125
Law
  affinity, 27
  Darcy's, 32
  Henry's, 33
  Hooke's, 67
  of motion, Newton's second, 102
  Stokes', 26, 38
Layer material strengths, 109
Layered
  geological strata, 121
  soil, 120
  soil, excavation, 122 (fig)
  strata, 120
Layout, curve, 112, 118
Leachate migration, 47
  landfills, 47
Learning curves, 128
Least radius of gyration, 66
Length, 92
  curve, comfort, 115
  cycle, 105
  development, 85
  development, masonry, 98
  spiral, 116
  spiral curve, 116

INDEX - L

Leveling, 110
  direct, 110 (fig)
  indirect, 110 (fig)
Lever, 64
Lifting equation, Niosh, 123
Lightweight aggregate factor,
    concrete, 75 (tbl)
Limit
  confidence, 16
  tests, Atterberg, 49
Line
  balance, 118 (fig)
  centroid of, 66
  energy grade, 22
  influence, 63
  straight, 12 (fig)
  waiting, 105
Linear
  algebra, 9
  deformation, 74
  momentum, 102
  motion, 101
  particle motion, 101
Liquefaction, 61
Load
  allowable pipe, 61
  buckling, 83
  combinations, 76
  distributed, 63
  eccentric, bolted connection, 72
  eccentric, rectangular footing, 53
  out-of-plane, 97, 99
  pressure, applied, 59
  pump shaft, 27
  thermal, 74
Loading
  axial member, eccentric, 69
  BOD, 42
  surcharge, 55
  thermal, 74
  treatment plant, 41
  weir, 43
Loads on buried pipe, 60
Local buckling, 93
Logarithm, 9
  identity, 9
Long-term deflection, 79
Loss
  combustion, 37
  contraction, 23
  energy, 22
  enlargement, 23
  head, 22
  heat, 46
  minor, 23
  prestress/pretension, 86
LRFD design, 88

M

M/M/1 single-server model, 105
M/M/s multi-server model, 105
Machine, hydraulic, 26
MACRS depreciation factor, 127 (tbl)
Management, project, 125
Manometer, 20 (fig)
Markings stake, 118
Marshall mix
  design test, 108 (fig)
  test, 108
Marston's formula, 60
Masonry
  columns, 100
  development length, 98
  properties of, 98
  wall, 98
Mass, 101
  conservation of, 22
  diagram, 118
  moment of inertia, 101
  sludge, 45
  -volume relationships, 49

Material
  strengths, layer, 109
  testing, 66, 120
Maximum
  allowable slope, 121
  allowable slope
      excavation, 121 (tbl)
  flame temperature, 37
  moment condition, 78 (fig)
  radius, spiral, 116
  steel area, 77
Mean
  geometric, 16
  harmonic, 16
  root, square, 16
Measurement
  distance, 110
  elevation, 110
  equal weight, 109
  flow weir, 29
  horizontal stadia, 110 (fig)
  unequal weight, 109
Measures of
  central tendency, 16
  dispersion, 16
Member tension, uniform thickness,
    unstaggered holes, 91 (fig)
Mensuration, 5
  of volume, 7
  three-dimensional, 7
  two-dimensional, 5
meq, 33
Meteorology, 31
Meter
  orifice, 24, 25 (fig)
  venturi, 24 (fig)
Methane production, 46
Method (*see also type*)
  average end area, 118
  depreciation, 126, 127
  double integration, 69
  dummy unit load, 74
  earned value, 125
  elastic, 97
  Greenshield's, 105
  instantaneous center of rotation, 97
  of coordinates, 111
  production output, 127
  rational, 31
  rational, peak runoff, 31
  service output, 127
  sinking fund, 127
  strain energy, 69
Meyerhof
  bearing capacity factor, 52 (tbl)
  factor, 52 (tbl)
Microbial growth, 41
Midspan deflections from
    prestressing, 86 (fig)
Migration
  leachate, 47
  leachate, landfills, 47
Milligram
  equivalent weight, 33
  per liter, 33
Million, parts per, 33
Minimum
  beam thickness, 80 (tbl)
  beam width, 77 (tbl)
  fillet weld size, 97 (tbl)
  steel area, 77
Minor loss, 23
Miscellaneous formulas, angles, 11
Mix design, Hveem, 108
Mixer, 39
  impeller, 39
Mixing, 13
  kinetics, 39
  physics, 39
Mixture, concrete (*see also type*), 75

Model
  M/M/1 single-server, 105
  M/M/s multi-server, 105
  plug flow, 44
  queuing, 105
  single-server queuing, 105
  stirred tank, 44
  stochastic critical path, 125
Modeling vehicle accidents, 107
Modified proctor test, 49
Modulus, 19
  bulk, 19
  of elasticity, 75, 98
  of elasticity, concrete, 75
  of rupture, concrete, 75
  section, 66
Moisture, 34
Molality, 33
Molarity, 33
Mole fraction, 33
Moment
  about a point, 63
  area method, 69
  available, versus unbraced
      length, 87 (fig)
  coefficients, 74
  component, 63
  distributed load, 63
  fixed-end, 73
  footing, overturning, 53 (fig)
  of equilibrium, 63
  of inertia, area, 66
  of inertia, cracked, 78, 79 (fig)
  of inertia, mass, 101
  of inertia, polar, 66
  overturning, 20
  resisting, 20
  ultimate plastic, 88
Momentum, linear, 102
Moody friction factor chart, 22 (fig)
Motion
  linear, 101
  uniform, 102
Motor
  electric, 27
  size, standard, 27
  speed, standard, 27
Moving average, forecast, 128
Multilane highway, 104
Multiplier, bearing capacity factor, 52 (tbl)
Municipal solid waste, 46
  incineration, 47

N

n-space, 9
National Research Council, 43
  equation, 43
Net
  area, 90
  area, effective, 91
  flow, 33
  flow, seepage, 33
  horizontal pressure dtribution,
      bulkhead, 58
  positive suction head, 27
Network
  activity-on-node, 125
  screening, HSM, 107
Neutralization, 34
Newton's second law of motion, 102
Niosh lifting equation, 123
No-stirrup condition, 81
Nominal moment strength, 78
Noise, 123
  dose, permissible, 124 (tbl)
  exposure level, 124 (tbl)
Nonrectangular channel, 30

Normal
    curve, area under standard, 14 (tbl)
    depth, 28
    distribution, 13
    force, 102
    form, 12
Normality, 33
NPSH, 27
NRC, 43
Number
    atomic, table, 34
    Froude, 26, 30
    Reynolds, 21
    structural, 109
    Weber, 26

O

Object height, 111
Occupancy
    group, 120 (tbl)
    requirements, 120
O'Connor, 41
Octagon properties, 6
Odor
    control, 40
    water, 40
Offset
    chord, 113 (fig)
    tangent, 112, 113 (fig)
One-tail confidence limit, 16
Open channel flow, 28
    parameters, 28
Optical density, 47
Order quantity, economic, 128
Orifice meter, 24, 25 (fig)
OSHA soil types, 120
Out-of-plane load, 99
Overconsolidated clay, 59 (fig)
Overturning moment, 20
    footing, 53 (fig)
Oxygen
    demand, biochemical, 42
    dissolved, wastewater, 41
    sag curve, 42 (fig)

P

Parabola, 13 (fig)
    formula, 5
Parabolic cable, 64 (fig)
Paraboloid of revolutions, 7
Parallel axis theorem, 66, 101
Parallelogram formula, 6
Parameters
    consolidation, 59, 60 (fig)
    fluid flow, 21
    open channel flow, 28
    sludge, 43
    soil, 49
    volume, 104
Parshall flume, 29
Particle size distribution, soil, 49
Parts per
    billion, 33
    million, 33
Passing sight distance, 115
Passive earth pressure, 54
Path
    critical, 125
    flow, 33
Pavement
    design, flexible, 107
    rigid, 109
    structural number, 109
Paving, 107
    equipment, 107
Peak
    -hourly factor, 104
    runoff, rational method, 31
Pedestrians, 105

Penetrometer test, cone, 49
Pentagon properties, 6
Per million, parts, 33
Permeability, 32
    tests, 49
Permeameter, 51 (fig)
Permissible noise dose, 124 (tbl)
pH, 34
Photogrammetry, 111
Physics
    mixing, 39
    sedimentation, 38
Pile, 56
    capacity, 56
    conical, 118 (fig)
    driver, 56
    group, 57
    shape, 118 (fig)
    volume, 117
    wedge, 118 (fig)
Pipe
    bend, 25 (fig)
    buried, loads on, 60
    culvert, simple, 23 (fig)
    flexible, 61
    load, allowable, 61
    rigid, 60
Pitot-static gauge, 23, 24 (fig)
Plain sedimentation basin, 43
Plane
    surface, pressure on, 20
    surveying, 109
Plastic
    design, 88
    moment, 88
Plate
    beam bearing, 88
    beam bearing, nomenclature, 90 (fig)
    column base, 93, 94 (fig)
    girder, 94 (fig), 95
    girder, flange, 95
    girder, web, 95
Plug flow, 44
    model, 44
pOH, 34
Point
    balance, 118 (fig)
    -bearing capacity, 57
    inflection, 69
    moment of force about, 63 (fig)
    pressure at, 59 (fig)
    -slope form, 12
    turning, 115
Poisson distribution, 13
Poisson's ratio, 66
Polar
    form, 12
    moment of inertia, 66
Pollutants, environmental, 47
Polygon
    formula, 6
    mensuration, 6
Polyhedra properties, 7
Porous filter, design, 32
Potential energy, 17, 21
    of a mass, 17
    scaling, 40
    stabilization, 40
Power, 16, 17
    aeration, 45
    conversion formulas, 17 (tbl)
    hydraulic, 26
    line clearance, 122
    line hazards, 122
    line, safety, 122
    pump, 26
ppb, 33
ppm, 33
Precipitator, electrostatic, 47
Preload, bolt, 72
Pressure
    at depth, 20

    at point, 59 (fig)
    culvert, 23
    distribution, 20
    earth, 53
    earth, active, 54
    energy, 21
    fluid, 20
    from applied load, 59
    hydrostatic, 20, 33
    on a plane surface, 20
    on a rectangular vertical plane
        surface, 20
    on buried pipe, 60
    passive, earth, 54
    soil, at-depth, 59
    soil, at-rest, 55
    vertical soil, 54
Prestress
    losses, 86
Prestressed concrete, 85
Prestressing, midspan deflections
       from, 86 (fig)
Primary
    consolidation, 60
    consolidation rate, 60
Principle, impulse-momentum, 103
    open system, 103
    work-energy, 17
Prismoidal formula, 118
Probability, 13
    and statistical analysis of data, 13
Problem
    CPM, 125
    fluid mixture, 13
    probabilistic, 128
Process efficiency, 44
Processes, wastewater treatment, 42
Processing, sludge, 43
Proctor test, 49
    modified, 49
Product vector
    cross, 10 (fig)
    dot, 10 (fig)
Production
    methane, 46
    output method, 127
Project
    management, 125
    ranking, 126
Properties
    concrete components, 75 (tbl)
    fluid, 19
    frame section, 82
    of area, 64
    of solid bodies, 101
    soil, 49
Pump
    centrifugal, 26
    efficiency, 26
    power, 26
    shaft loading, 27
    similarity, 27
Purification
    dilution, 42
    response to, dilution, 42
Pyramid, rectangular, 118 (fig)

Q

Quadratic equation, 9
    root, 9
Quality
    break-even, 128 (fig)
    characterized rock mass, 51
    wastewater, 41
    water supply, 37
Quantity
    economic order, 128
    sludge, 39, 45
    wastewater, 41
Queuing model, 105

R

Radian, 10
Radical, 9
Radius
  hydraulic, 21, 28
  of gyration, 66, 101
Raft on
  clay, 53
  sand, 53
Railroad
  line, superelevation, 114
  rolling stock, 106
Rainfall intensity, 31
Rankine theory, 53
Ranking
  economic, 126
  projects, 126
Rate
  braking, 107
  consolidation, 60
  deceleration, 107
  primary consolidation, 60
  recirculation, 45
  return, 45
Ratio
  available compressive stress versus
    slenderness, 92 (fig)
  benefit-cost, 126
  depth-thickness, 94
  food microorganism, 42
  Poisson's, 66
  recycle, 45
  scale, 26
  slenderness, 92
  volume-capacity, 104
  width-thickness, 95
Rational method, 31
  peak runoff, 31
Reaction
  combustion, ideal, 36 (tbl)
  enthalpy of, 34
  influence line, 63
  kinetics, reversible, 34
Reactions, combustion, 36
Recirculation rate, 45
Rectangular
  channel, 29, 30
  footing, eccentric load, 53
  pyramid, 118 (fig)
  vertical plane surface, 20
  weir, 29
Recycle ratio, 45
Reduction value, 91 (tbl)
  welded connections, 92 (tbl)
Regeneration, ion, 40
  exchange resin, 40
Reinforced concrete, 81
Reinforcement, anchorage, shear, 81
Reinforcing steel, pavement, 109
Relationship
  mass-volume, 49
  speed, flow, and density, 104
  weight-volume, 107
Relative torsional stiffness, 81
Reliability, application, 16
Remediation, environmental, 47
Reoxygenation, 41
Reservoir routing, 32
Resilience, 67
Resin, ion, 40
Resistance, rolling, 102
Resisting moment, 20, 21
Resources, water, 19
Restitution, coefficient of, 104
Resultant distribution, base, 56 (fig)
Retaining wall, 83
  analysis, 56
  analysis, cantilever, 56
  design, 56, 83
  design, cantilever, 56
  rigid, 53
Retention, specific, 32

Return rate, 45
Reversible reaction, 34
  kinetics, 34
Reynolds number, 21
Right
  circular cone, formula, 7
  circular cylinder, 7
  triangle, 10
Rigid
  body, 103
  pavement, 109
  pavement design, AASHTO method, 109
  pipe, 60
  retaining wall, 53
Rigidity, 67
Rivet, 71
  connection, 71
RMS, 16
Road capacity, 104
Roadway banking, 103
Rock, stable, excavating, 121 (tbl)
Rod, grade, 118
Rolling
  resistance, 102
  resistance, wheel, 103 (fig)
  stock, railroad, 106
Root
  quadratic equation, 9
  mean square, 16
Rope, wire, 73
Rotation, center of, 97
Routing, reservoir, 32
Rule
  Simpson's, 12
  trapezoidal, 12
Rules for
  exponents, 9
  radicals, 9
Runoff, peak, 31
Rupture, modulus of, 75
  concrete, 75

S

Safety, 120
  electrical, 123 (tbl)
  job site, 120
Sag
  curve, 115
  curve, oxygen, 42 (fig)
  oxygen, 42
Salinity, 37
  irrigation water, 37
Salvage value, replacement study, 126
Sand
  bearing capacity, 52
  braced cut, 57
  cuts in, 57 (fig)
  raft on, 53
  stability of braced excavations, 58
  stability, cut, 58
Scale ratio, 26
Scheduling, 125
Screen, gravel, 32
Second-order
  analysis, 82
  effects, 94
Secondary consolidation, 60
Section
  compact, 87
  modulus, 66
Sector, 5
Sedimentation, 38
  basin, plain, 43
  clarification, plain, 43
  physics, 38
  tank, 39
Seeded BOD, 42
Seepage, 33
  flow net, 33
  velocity, 32
Selection of flexural reinforcement, 85

Sensitivity, 51
Separator, cyclone, 47
Series, galvanic, 34 (tbl)
Service
  load, 76
  output method, 127
Settling velocity, 38
Shaft
  design, 72
  loading, pump, 27
Shallow
  foundation, 51
  water table, 52
Shape
  mensuration, 5
  pile, 118 (fig)
Shear
  ACI coefficient, 74 (tbl)
  cantilever wall, 58 (fig)
  critical section, one-way, 84 (fig)
  critical section, two-way, 85 (fig)
  in prestressed sections, 86
  lag reduction coefficient, bolted
    connection, 91 (tbl)
  lag reduction coefficient, welded
    connection, 92 (tbl)
  reinforcement, 80
  reinforcement design, 81
  reinforcement, anchorage, 81
  reinforcement, limitations, 80
  strength, 88, 95
  strength, block, 91
  strength, steel beam, 88
  stress, beam, 68
  stress, rectangular beam, 68 (fig)
  test, direct, 51
  ultimate, 88
Shell, torsion, 72
Shield configuration, 121 (fig)
Shrinkage, 85, 117
  factor, 117
Sight distance, 115
  passing, 115
  stopping, 114
Sign convention, 68 (fig)
Signal, 105
  timing, 105
Signalized intersection, 105
Similarity, 26
  pump, 27
Simple
  framing connection, 97 (fig)
  pipe culvert, 23 (fig)
Simpson's rule, 12
Sine, 10
Single
  -payment equivalence, 125
  -server queuing model, 105
Sinking fund method, 127
Size factor, 85
Skidding distance, 107
Skin-friction capacity, 57
Slab
  beam, two-way, 82 (fig)
  beams, factored moments in, 81
  concrete, effective width, 96
  deflection, 82
  design, 81
  width, effective, composite member, 86
Slack, 125
Slenderness ratio, 91, 92
Slip, 27
Slope, 12
  configuration, 121 (fig)
  geometric, 28
  -intercept form, 12
  maximum allowable, 121 (tbl)
  maximum allowable,
    excavation, 121 (tbl)
  stability, clay, 60
  stability, Taylor, 60 (fig)
  staking, 118

Sludge, 43
  activated, 43
  dewatering, 46
  mass, 45
  parameters, 43
  processing, 43
  quantity, 39, 45
  volume, 45
  volume index, 43
Small angle approximation, 11
Smoke, 47
Soft water, 40
Softening
  ion exchange, 40
  water, 40
Soil (see also type)
  classification, 49, 120
  classification, AASHTO, 49
  indexing formula, 50 (tbl)
  layered, 120
  organic, 59
  parameters, 49
  particle size distribution, 49
  pressure at-depth, 59
  pressure, at-rest, 55
  pressure, vertical, 54
  properties, 49
  surcharge, 55
  testing, 49
  topics, special, 59
  type, 120
  type A, 120
  type B, 121
  type C, 121
  type, geotechnical quality, 121 (fig)
  types, OSHA, 120
Solid
  body, 101
  body, properties of, 101
  waste incineration, municipal, 47
  waste, municipal, 46
Soluble BOD, 43
  escaping treatment, 43
Solution of gases in liquids, 33
Sound, speed of, 19
Spacing, stirrup, 81
Special soil topics, 59
Specific
  capacity, 32
  energy, 22, 29
  gravity, 19
  retention, 32
  speed, 27
  speed, suction, 27
  speed, turbine, 28
  volume, 19
  weight, 19
  yield, 32
Specification, direction, 110
Speed, 101
  of sound, 19
  relationship, 104
  specific, 27
  suction specific, 27
  turbine specific, 28
Sphere
  drag on, 25
  formula, 7
  terminal velocity, 25
Spherical
  segment, 7
  tank, 71
Spillway, 29
Spiral
  column, 82
  curve, 111, 115, 116 (fig)
  curve length, 116
  length, 116
  maximum radius, 116
Splitting tensile strength, 75
Spring, 72
  energy, 17

Square, root mean, 16
Stability
  braced excavation, clay, 58
  braced excavation, sand, 58
  sand, cut, 58
  slope, clay, 60
  Taylor slope, 60 (fig)
Stable rock, excavating, 121 (tbl)
Stadia measurement, horizontal, 110 (fig)
Stake
  marking abbreviation, 119 (tbl)
  markings, 118
Staking, 118
  slope, 118
Standard
  deviation, 16
  motor size, 27
  motor speed, 27
  normal curve, area under, 14 (tbl)
Start
  earliest, 125
  latest, 125
Statics, fluid, 20
Stationing, curve, 112
Statistics, 13
Statutory depreciation, 127
Steel
  beam, analysis flowchart, 89 (fig)
  beams, 87
  beams, minimum cost, 77
  column, design, 93
  connector, 96
  cover, 77
  maximum, area, 77
  minimum, area, 77
  reinforcing, pavement, 109
  reinforcing, property, 74
  structural, property, 87 (tbl)
  temperature, 81
  wheel, 106
Stiffener
  bearing, 96 (fig)
  interior, 95
  intermediate, 95 (fig)
Stiffness, 67
Stirred tank, 44
  model, 44
Stirrup spacing, 81
Stirrups, none required, 81
Stochastic critical path model, 125
Stoichiometric air, 36
Stokes' law, 26, 38
Stopping
  distance, 106, 114
  sight distance, 114
Straight line, 12 (fig)
  depreciation, 127
Strain energy, 67, 69
  method, 69
  bending moment, 69
  total, 67
Strata, layered, 121
  geological, 121
Stream
  confined, 25
  coordinates, fluid, 23 (fig)
  in pipe bend, confined, 25
Strength
  available, bearing, 96
  block shear, 91
  compressive, 74
  design, 76
  flexural, 87
  flexural, available, 96
  of materials, 67
  provisions, ACI, 86
  shear, 88, 95
  steel beam, 87, 94, 95
  tensile, 90
  test, unconfined compressive, 51

Stress
  bending, 68
  combined, 68 (fig)
  concentration, 68
  cylinder, thick-walled, 71 (tbl)
  effective, 55
  in beams, shear, 68
  tensile, axial, 90
  test, triaxial, 51
  true, 66
Stripping, air, 47
Structural
  analysis, 73, 74
  number, 109
  number, pavement, 109
  steel, 87
  steel, property, 87 (tbl)
Structure, composite, 69
Strut, 57 (fig)
Student's t-distribution, 13
  values of t, 15 (tbl)
Suction
  head, net positive, 27
  specific speed, 27
Sudden
  contraction, 23
  enlargement, 23
Sum-of-the-years' digits depreciation, 127
Superelevation, 114
  railroad line, 114
  transition, 114
Supply
  quality, water, 37
  water, 37
Support, lateral, 87
Surcharge
  loading, 55
  soil, 55
Surface, tension force, 26
Surveying, plane, 109
SVI, 43
Swell, 117
  factor, 117
Symbol
  cash flow, 126
  chemical, 34
System curve, 27

T
t-distribution, student's, 13
T-beam, 80
Table
  of elements, 34
  water, 53
Tacheometry, 110
Tangent, 10
  offset, 112, 112 (fig)
  offset, curve layout, 113
Tank
  aeration, 44
  discharge, 23 (fig)
  model, stirred, 44
  sedimentation, 39
  spherical, 71
  stirred, 44
  thin-walled, 70
  thin-walled spherical, 71
Taping, 110
Taste
  control, 40
  water, 40
Tax, income, 127
Taylor slope stability, 60 (fig)
Temperature
  combustion, 37
  flame, 37
  maximum, flame, 37
  steel, 81
Temporary
  excavation, 57
  traffic control zone, 106

Tendency, measures of central, 16
Tensile
    strength, 90
    strength, splitting, 75
    stress, 90
    test, 66
Tension
    connection, 96
    connection, concentric, 97
    force, surface, 26
    member, steel, 90
    member, uniform thickness, unstaggered
        holes, 91 (fig)
Terminal velocity, 25
    sphere, 25
Terzaghi factor, 52 (tbl)
Test (see also type)
    Atterberg limit, 49
    California bearing, 51
    CBR, 51
    cone penetrometer, 49
    consolidation, 51
    direct shear, 51
    Marshall mix, 108
    Marshall mix design, 108 (fig)
    modified proctor, 49
    permeability, 49
    proctor, 49
    tensile, 66
    torsion, 67
    triaxial stress, 51
    unconfined compressive strength, 51
Testing
    material, 66, 120
    soil, 49
    water quality, 37
Theis equation, 33
Theorem, parallel axis, 66, 101
Theory
    Coulomb, 54
    Rankine, 53
Thermal
    deformation, 67
    load, 74
Thick-walled cylinder, 71
    stress in, 71 (fig)
Thiem equation, 32
Thin-walled tank
    cylindrical, 70
    spherical, 71
    stress in, 71 (fig)
Three
    -dimensional mensuration, 7
    -moment equation, 73
Tied column, 82
Time of concentration, 31
Timing
    signal, 105
    traffic-activated, 105
Topic, special soil, 59
Torque, bolt, 72
Torricelli's equation, 23
Torsion
    shell, 72
    test, 67
Torus formula, 7
Total strain energy, 67
Toughness, 67
Traffic, 109
    accident analysis, 107
    -activated timing, 105
    control zone, temporary, 106
    design, 109
Transcendental function, circular, 10
Transfer
    heat, 46
    of force at column base, 85
Transition, superelevation, 114
Transmissivity, 32
Transportation, 101
Trapezoid formula, 6

Trapezoidal
    channel, 28, 29
    cross section, 29 (fig)
    rule, 12
Traverse, 111
    area, 111
    balanced, 111
Treatment
    escaping, 43
    plant loading, 41
    wastewater, 42
    water supply, 38
Trenching, 121
    safety, 121
Triangle
    formula, 5
    general, 11 (fig)
    right, 10
Triaxial stress test, 51
Trickling filter, 43
Trigonometric
    functions in a unit circle, 10 (fig)
    identity, 11
Trigonometry, 10
Trip generation, 104
Truck factor, 109
True
    strain, 66
    stress, 66
Truss
    deflection, 69
    determinate, 64
    indeterminate, 73
TSS equation, 41
Turbine specific speed, 28
Turbulent
    energy loss, 22
    friction loss, 22
    head loss, 22
Turning point, 115
Two
    -angle formula, 11
    -dimensional mensuration, 5
    -dimensional space, vector in, 9 (fig)
    -lane highways, 104
    -point form, 12
    -tail confidence limit, 16
    -way slab beam, 82 (fig)
Type
    A soil, 120
    B soil, 121
    C soil, 121

U

Ultimate
    plastic moment, 88
    shear, 88
    strength design, 70
Unbraced
    column, 82, 83
    length versus available moment, 87 (fig)
Unconfined compressive strength test, 51
Uniform
    acceleration, 102
    acceleration formulas, 102 (tbl)
    flow, equations, 28
    motion, 102
Unit
    hydrograph, 31
    of concentration, 33
    vector, 10
    vector, Cartesian, 10 (fig)
    weight coefficient, value of, 76 (tbl)
Unsteady flow, 33
Unsymmetrical bending, beam, 88

V

Value
    book, 127
    chemistry coefficient, 76 (tbl)
    method, earned, 125
    of t, student's t-distribution, 15 (tbl)
    of z, 16 (tbl)
Valve, 23
    flow coefficient, 23
Variation, coefficient of, 16
Varved clay, 59
Vector, 9
    cross product, 10 (fig)
    dot product, 10 (fig)
    in n-space, 9
    in two-dimensional space, 9 (fig)
    unit, 10
Vehicle
    accidents, modeling, 107
    dynamics, 106
Velocity
    discharge, 23, 32
    Hazen-Williams, 28
    seepage, 32
    settling, 38
    terminal, 25
    terminal, sphere, 25
Velz equation, 43
Venturi meter, 24 (fig)
Vertical
    curve, 114
    curve length, 115
    curve, obstruction, 115 (fig)
    curve, symmetrical parabolic, 114 (fig)
    curve, through point, 115
    force, elements, 56 (fig)
    plane surface, hydrostatic pressure
        on, 20 (fig)
    plane surface, rectangular, 20
    soil pressure, 54
Vesic factor, 52 (tbl)
Virtual work, 69
Viscosity, 19
    absolute, 19
    kinematic, 19
Viscous force, 26
Volume
    -capacity ratio, 104
    index, sludge, 43
    mensuration, 7
    mensuration of three-dimensional, 7
    parameter, 104
    pile, 117
    sludge, 45
    specific, 19

W

Waiting line, 105
Walkway, 105
Wall
    bearing, 83
    design, ASD, 99
    footing, 84
    retaining, 83
    retaining, analysis, 56
    retaining, design, 56, 83
    rigid retaining, 53
    shear, cantilever, 58 (fig)
Waste, municipal solid, 46
Wastewater, 41
    quality, 41
    quantity, 41
    treatment, 42
    treatment, equipment, 42
    treatment, processes, 42
Water
    demand, 40
    hammer, 25 (fig)
    hammer, one-way, 25
    hammer, round trip, 25

INDEX - W

horsepower, 26
odor, 40
properties, 75
quality conversions, 37
quality testing, 37
resources, 19
softening, 40
supply, 37
supply demand, 40
supply distribution, 38
supply quality, 37
supply treatment, 38
table, 53
table, bearing capacity, 53
table, shallow, 52
taste, 40
Web
 flanged beam, 68 (fig)
 force, concentrated, 88
 plate girder, 95
 yielding calculation,
  nomenclature, 90 (fig)
Weber number, 26
Webster's equation, 105
Wedge pile, 118 (fig)
Weight, 101
 atomic, 35 (tbl)
 equivalent, 33
 milligram equivalent, 33
 rectangular, 29
 specific, 19
 -volume relationships, 107
Weir, 29
 broad crested, 29
 flow measurement, 29
 loading, 43
 rectangular, 29
Weld
 fillet, 72, 97 (fig)
 size, 97 (tbl)
Welded connection, combined shear
  and bending, 97 (fig)
Welding, 97
Well drawdown, 32
 in aquifers, 32
Wheel, steel, 106
Width
 beam, minimum, 77 (tbl)
 effective, 96
 effective slab, composite member, 86
 -thickness ratio, 93 (tbl), 95
Wire rope, 73
Work, 16, 74
 -energy principle, 17
 of a constant force, 16 (fig)
 virtual, 69

Y

Yield, specific, 32

Z

z-values for confidence level, 16 (tbl)
Zone of influence, 59